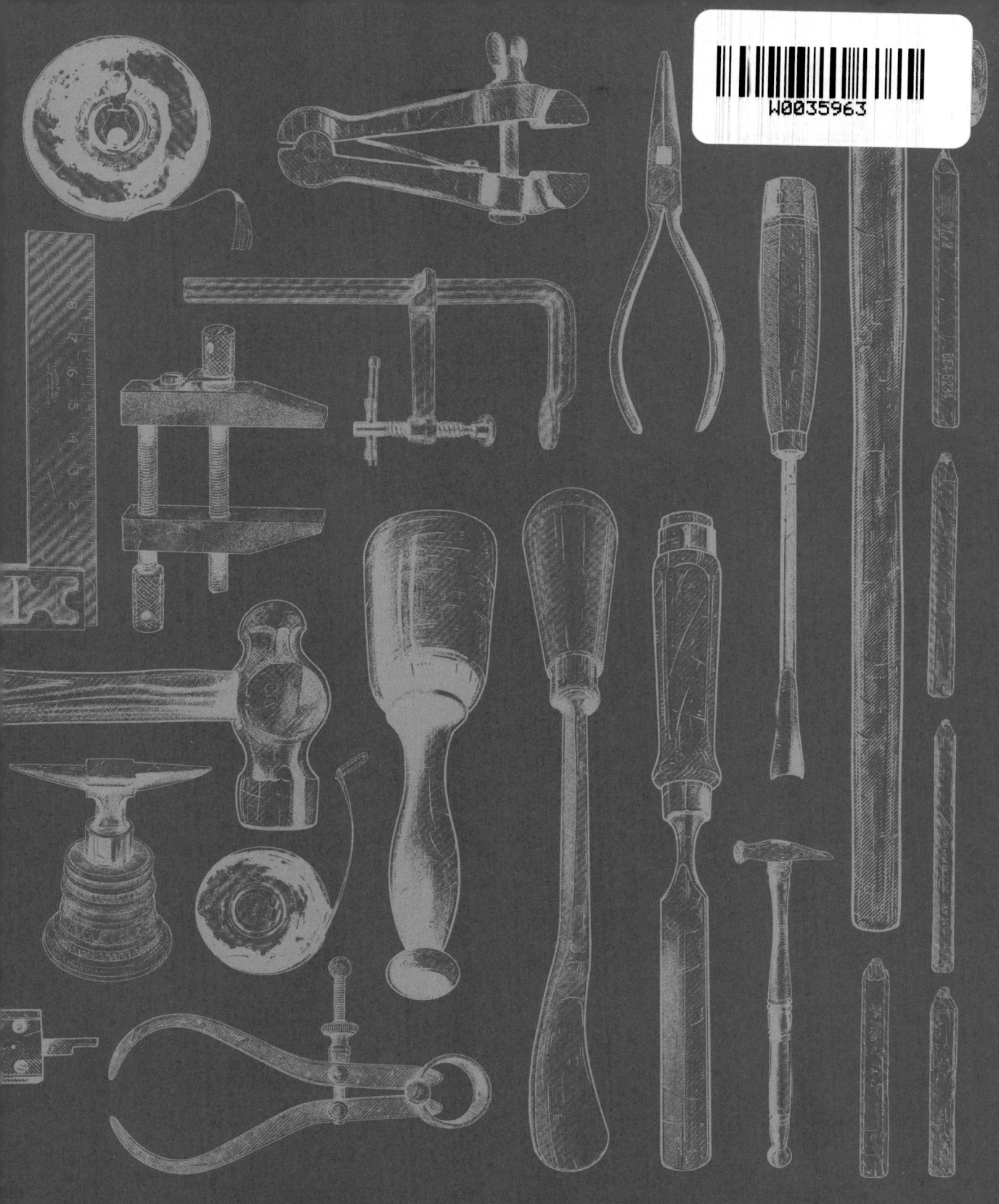

TOOLS

A VISUAL HISTORY

DOMINIC CHINEA

Illustrated by **LEE JOHN PHILLIPS**

Editor Emma Bastow
Project and Jacket Designer Eoghan O'Brien
Designer James McKeag
Production Editor Siu Yin Chan
Senior Production Controller Louise Minihane
Senior Acquisitions Editor Pete Jorgensen
Managing Art Editor Jo Connor
Publishing Director Mark Searle

Text Dominic Chinea and Nathan Joyce
Illustrations Lee John Phillips

DK would like to thank Katie Crous for proofreading and Vanessa Bird for indexing.

First published in Great Britain in 2022 by
Dorling Kindersley Limited
DK, One Embassy Gardens, 8 Viaduct Gardens, London SW11 7BW

The authorised representative in the EEA is
Dorling Kindersley Verlag GmbH. Arnulfstr. 124,
80636 Munich, Germany

Page design copyright © 2022 Dorling Kindersley Limited
A Penguin Random House Company
10 9 8 7 6 5 4 3
003–329007–Oct/2022

Text copyright © 2022 Dominic Chinea

A CIP catalogue record for this book is available from the British Library.
ISBN: 978-0-2415-6196-6

Printed and bound in China
For the curious
www.dk.com

This book was made with Forest Stewardship Council™ certified paper—one small step in DK's commitment to a sustainable future. For more information go to www.dk.com/our-green-pledge

CONTENTS

INTRODUCTION

I get asked a lot what my favourite tool is, and it's an impossible question to answer because it depends on what I'm doing at that moment in time. Some days, when I'm staring at a nut that used to be hexagonal and it's so rusty that it won't undo, my favourite tool is the old, misshapen spanner that wiggles it free. I might not use that spanner for another three years, but wow, I'm grateful that I bought it at that car boot sale 15 years ago! I think that's part of the reason why I like collecting old hand tools, because at some point they'll be the only thing to help you out of a tight spot. I spend a lot of time looking for tools and bits of equipment in car boot sales. Sometimes I'll have a vague plan of what
I might do with it, and sometimes not. I buy almost everything second-hand, but on the odd occasion I do buy new tools, they're usually made by craftspeople who are continuing or have revived a centuries-old tradition.

Sometimes I come across a hand tool displaying such stunning craftsmanship that I just marvel at the skill and dedication that someone has committed to their craft. I love that feeling of finding the diamond in the rough. I see myself as a kind of custodian of tools; I'll look after them, and eventually, when the time is right, I'll give them to someone who I know will love and

cherish them along with any wisdom I've gleaned on the way. That's what heritage craft is all about.

I've found myself in a uniquely privileged situation on *The Repair Shop,* in a job that's impossible to train for. I'm learning all the time as I'm going, but a lot of the work isn't just about practical skills. Much of it comes down to putting yourself into someone else's shoes, imagining how they tackled something and wondering which tools they used. I go into my workshop each day excited about what's going to land on my workbench. And when it does, I always feel lucky to have been trusted to work on that precious item.

A question I get asked often is if I've ever had something I can't fix. Things go wrong a lot – the very nature of *The Repair Shop* is tackling difficult-to-repair objects, but you learn to work out how to solve each problem in turn. Pete, our musical-instrument repairer, always jokes that being a "professional" at something means knowing how to get yourself out of trouble (although he says it a little stronger than that). He's right, because the buck stops with you, and it's not like we can blame our tools!

One of the items I've worked on at *The Repair Shop* that I'll remember for the rest of my life was an old hand tool: a hammer with a loose head. The chap who brought it in told me that his beloved grandad had used that

Over time, tools become your friends – you spend so much time with them, they help you out and you come to rely on them.

hammer for everything, including building his house, and his grandson wouldn't ever give up on it because he never wanted to lose that connection to his grandad. The handle was weirdly twisted and bent, and the end of the handle was completely flat and full of bits of ply. Little screws and nails were only just holding it together. It had clearly had an amazing life, so I didn't want to be the one to say it needs a new handle. It wasn't for me to give up on it. So I gave it a go, steaming it straight, then clamping it in three directions. I've never been more nervous working on an object. The handle kept bending and bending, and I was terrified it might just snap. But I convinced myself that it was alright if it did, because at least I'd tried. So some of the chap's grandad's life lessons were rubbing off on me as well.

In the end, I filed down the head so it ended up sitting a bit lower on the handle. I cleaned the head but didn't buff it up to a mirror polish because I didn't want it to become just any other hammer. I wondered about adding his grandad's initials to it for ages. It was a hard decision because once you commit to it, there's no going back. But I did commit. The time, respect and determination paid off (and luck probably paid its part too) and the chap was hugely moved. And that old hammer will get used again. Who knows what it'll be used to make in the future, but it's already got quite a tale to tell.

I love reviving incredible crafting techniques, tools and pieces of equipment that are in danger of falling by the wayside. But the one closest to my heart is the wheeling machine (also known as an English wheel). An amazing piece of equipment, it has a huge cast-iron frame leading to two ground and polished steel wheels that you feed sheet metal back and forth through to produce curves. Many coachbuilders (craftspeople who make the bodies of cars, buses and railway carriages) in the 1920s and 1930s used them, but only one actually made their own wheeling machine: Ranalah, and they produced the Rolls Royce of wheeling machines, which is fitting, seeing as Ranalah helped produce Rolls Royces.

Around 2m (7ft) tall, 1.2m (4ft) wide, weighing just under 1,000kg (1 ton) and shaped like a giant "C", the Ranalah English wheel is an icon. I fell in love with that company and the wheel so much that I ended up buying the rights to produce the wheels again under the Ranalah name. But that was only half the battle. I had to actually find an original Ranalah so that I could use it as the template to cast the first Ranalah English wheel made in 90 years. Now, I want to make these English wheels available to as many people as possible, from engineering firms keen to preserve our industrial heritage to younger people who want to take up the craft.

In this book, you'll find over 140 hand tools, many of which have remarkable stories of their own. Some of the tools will take you back to your dad's toolbox, nan's sewing box, your grandparents' shed or maybe even your first woodworking class at school.

I love the unique connection old tools give you to the past; also, tools back then were usually made with care, treated with respect and built to last. Continuing that tradition is what I do for a living, and I wouldn't have it any other way.

DOMINIC CHINEA

CHAPTER I
MEASURING & MARKING

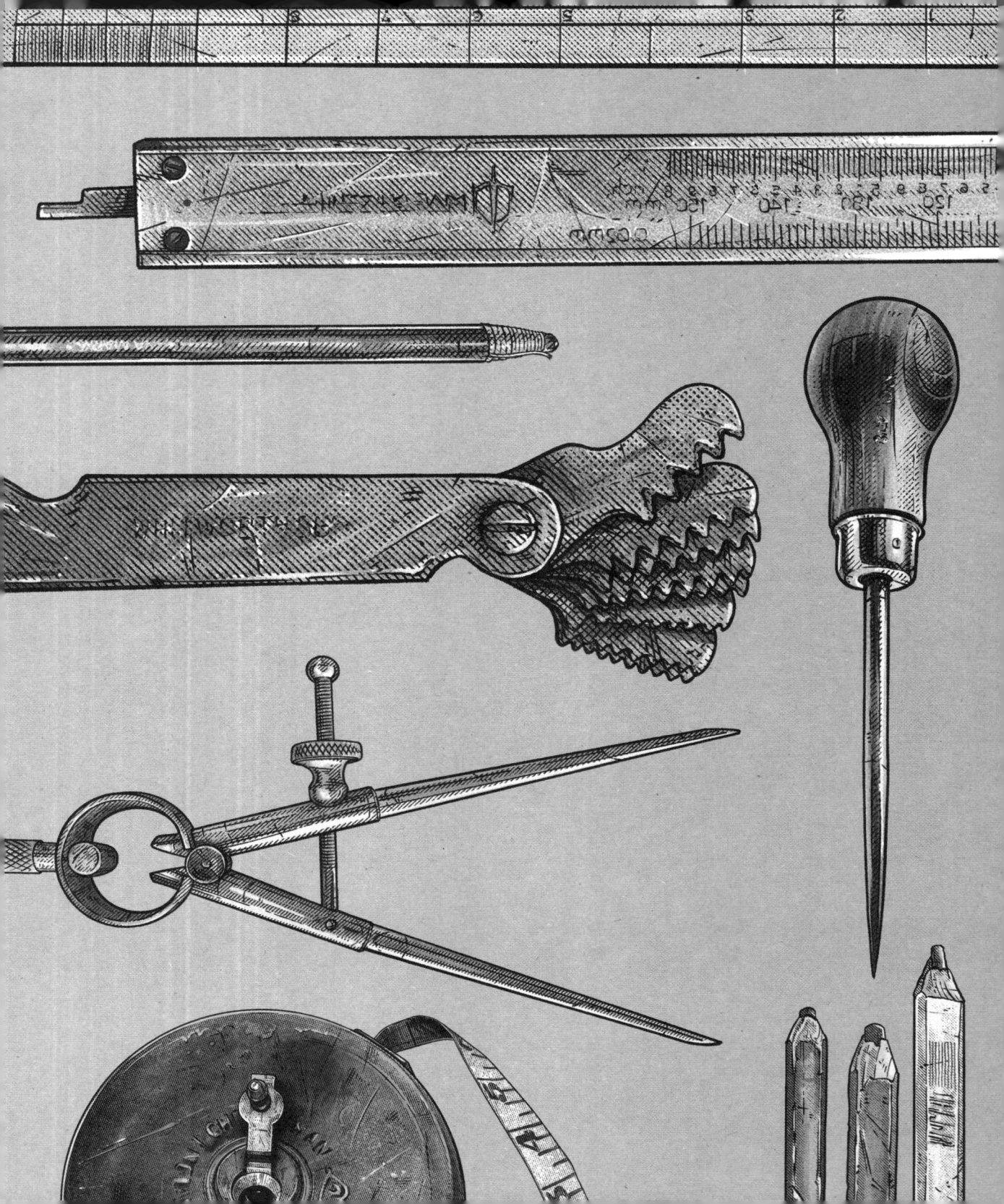

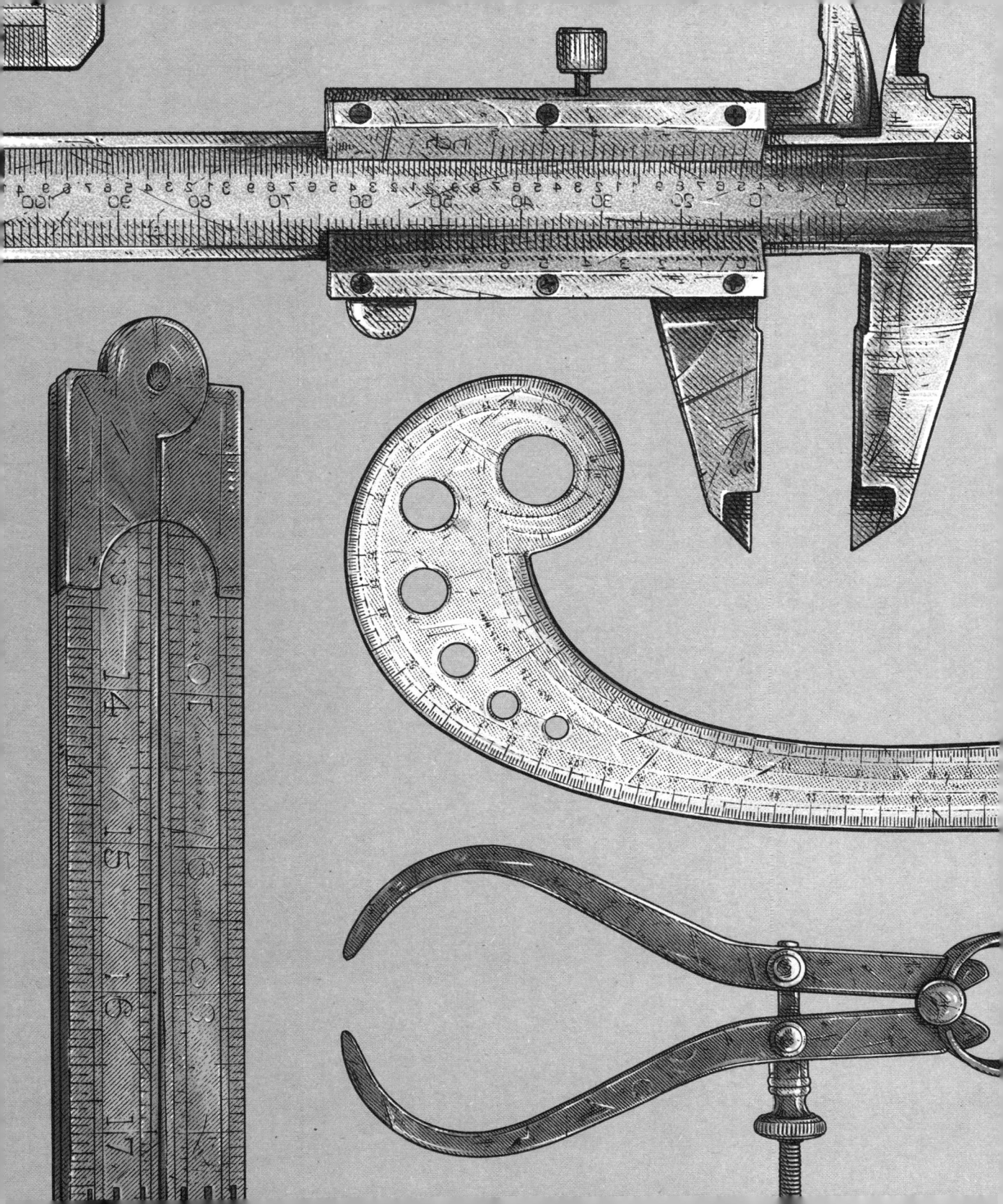

CARPENTER'S RULER

A carpenter's ruler is a folding ruler with metal (usually brass) ends and a central arched joint. Many older carpenter's rulers were made of boxwood, which starts off a light cream or yellow colour then develops a beautiful honey or toffee patina over time. The only trouble with boxwood is that it warps. That's why the brass edges are fitted – to hold the wood in position.

Carpenter's rulers are designed to be both portable and flexible and range from just a few centimetres/inches long to 2-metre/yard-long ones. However long they are, they're all designed to fold down and fit neatly in your pocket. Another big advantage is that they can be used in tight spaces where a larger ruler just won't work. They're not as accurate as machinist's rulers (see opposite), as they don't have the same number of measuring increments and the joints can slip slightly over time.

Carpenter's rulers were good enough for the Romans, though. A folding ruler, made of two rectangular bronze bars that were hinged together, was even found in the ruins of Pompeii. Together, the two pieces measured 29.5cm (11½in), or a Roman foot, roughly 1cm (½in) less than the modern measurement for a foot. In the Georgian and Victorian eras, carpenter's rulers were sometimes carved from bone or ivory. Birmingham seems to have been the place to be if you were a ruler maker just as the Industrial Revolution was heating up. You can see why it earned its name the "city of a thousand trades" around then. By the 1780s, there were nine ruler makers in and around the city, including F.B. Cox and John Rabone & Sons, who made some of the finest ones around. If you look carefully, you can still pick up rulers bearing these names at a car boot sale.

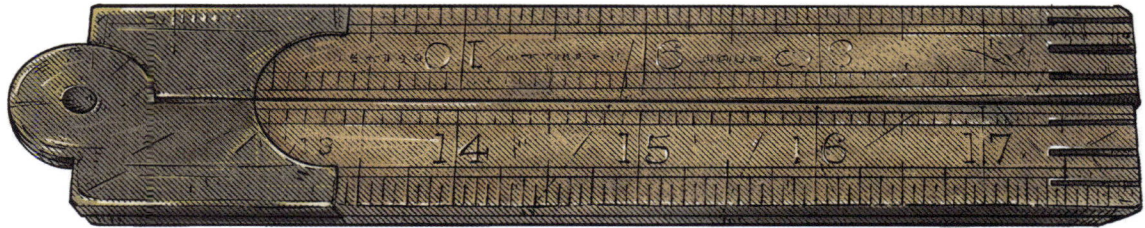

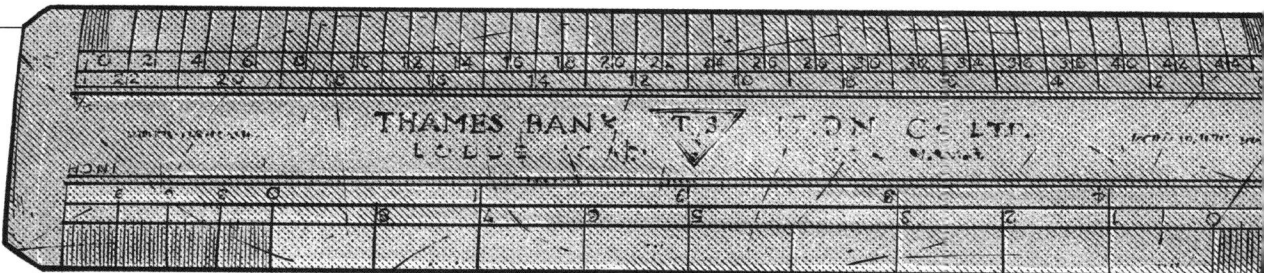

Tool No. **2**

MACHINIST'S RULER

A ruler has two simple but important functions: to draw a straight line and to measure distances. A machinist's ruler is made from stainless steel and features increments in both imperial and metric units that are etched onto both sides. It's a far cry from the bog-standard, apparently shatter-resistant plastic ruler you started the new school term with.

The first thing you notice on a machinist's ruler is the higher number of increments between inches and centimetres – they're usually marked every 0.5mm and every ¹⁄₆₄in, which ensures a high degree of accuracy. And you'll need that if you're a mechanic, engineer or architect and you're working with detailed pieces of equipment or technical drawings. Machinist's rulers are made in a number of sizes, from 150mm (6in) up to a pretty massive 2,000mm (80in).

The steel used to make machinist's rulers usually undergoes a process of hardening, which involves heating it gradually to a very high temperature of around 800–900°C (1,450–1,650°F), then cooling it immediately in water or oil. Then the steel is tempered, by heating it again to a relatively low temperature of around 200–250°C (400–500°F), before cooling it again. Tempering makes the steel slightly less hard, while also making it less brittle. It's all about getting the balance right between hardness and toughness. Think about it as giving you both strength and endurance.

Some machinist's rulers are very thin and are designed to be really flexible, so you can use them to measure a curve. I sometimes use them to draw an arc between two points, to give a nice flowing line, like you would with a French Curve ruler (see page 12).

Tool No. **3**

PATTERN-MAKING RULER

To make an item of clothing that's tailored to someone's shape, you first need to measure various areas of the body and draw up a paper template. This is more complicated than it sounds, because every item of clothing is actually several different pieces of fabric that are sewn together. To draw up the template – known as pattern drafting – you'll need specialized rulers.

One specialized ruler is called a French curve, which looks a bit like a stretched comma. It's made of several different shaped curves and doesn't have any straight edges. A French curve is used to make armholes, waistlines, lapels and necklines, and is also really useful for making pattern alterations.

It's not known exactly when French curve rulers appeared, but they've been around since at least the middle of the 19th century, and have also been used by mathematicians, engineers and architects when drafting technical drawings. A book published in 1849, with just about the longest title I've ever seen (it starts *A Rudimentary Treatise on Masonry and Stonecutting,* but that's only a quarter of it), mentions "The curved rulers manufactured in Paris of thin veneer sold under the name *French curves,* are very useful for drawing in between points previously determined small portions of elliptical or other curves."

The modern ones used for sewing are made of transparent plastic or aluminium, but you can still pick up beautiful old wooden ones if you hunt around a bit.

Tool Nos. **4–6**

CALIPERS & DIVIDERS

A pair of calipers is a tool used to measure the dimensions of objects that can't be measured easily with a ruler. Confusingly, there are several different types of caliper, each of which does different things. Inside calipers, outside calipers and oddleg calipers all look like variations of a pair of compasses, but Vernier calipers (see page 15) are the tool that most people think of when someone says, "Hand me the calipers".

Vernier calipers have one fixed jaw and one adjustable one that slides, and they're named for the French mathematician Pierre Vernier who came up with them in 1631. The instrument uses two scales, one main scale that looks like a regular ruler and a second scale that slides parallel to the main scale. This second scale, known as the Vernier scale, lets you take much more precise readings. The basic design for this

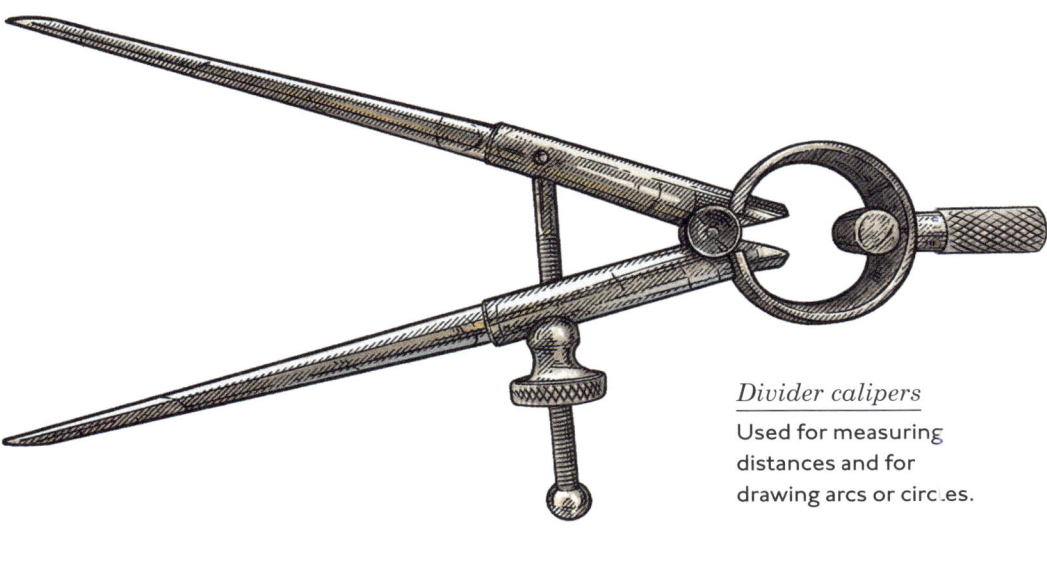

Divider calipers
Used for measuring distances and for drawing arcs or circles.

type of caliper dates back to at least the 6th century BCE. A pair, made of wood and featuring one fixed and one movable jaw, were found in a Greek (or possibly Etruscan) shipwreck off the coast of the Isola del Giglio, west of Tuscany, Italy, and were discovered by marine archaeologists in the 1980s.

An inside caliper, with legs that curve outwards, measures the internal size of something. An outside caliper (see opposite), with legs that bow outwards before coming together at the bottom, measures the external size of an object. An oddleg caliper (also known as Jenny calipers), has one straight leg and one bent leg, which is used to draw lines parallel to the edge of an object. Divider calipers (see page 13), better known as a pair of compasses, are used for measuring distances and for drawing arcs or circles. They all have adjustable screws for fixing the legs in position.

I've got a nice old pair of outside calipers, which are really handy to measure curved surfaces, such as the width of pipes. I've also got a couple of beautiful old dividers that look like something Christopher Columbus would keep in a nice wooden box before getting to work planning an epic voyage.

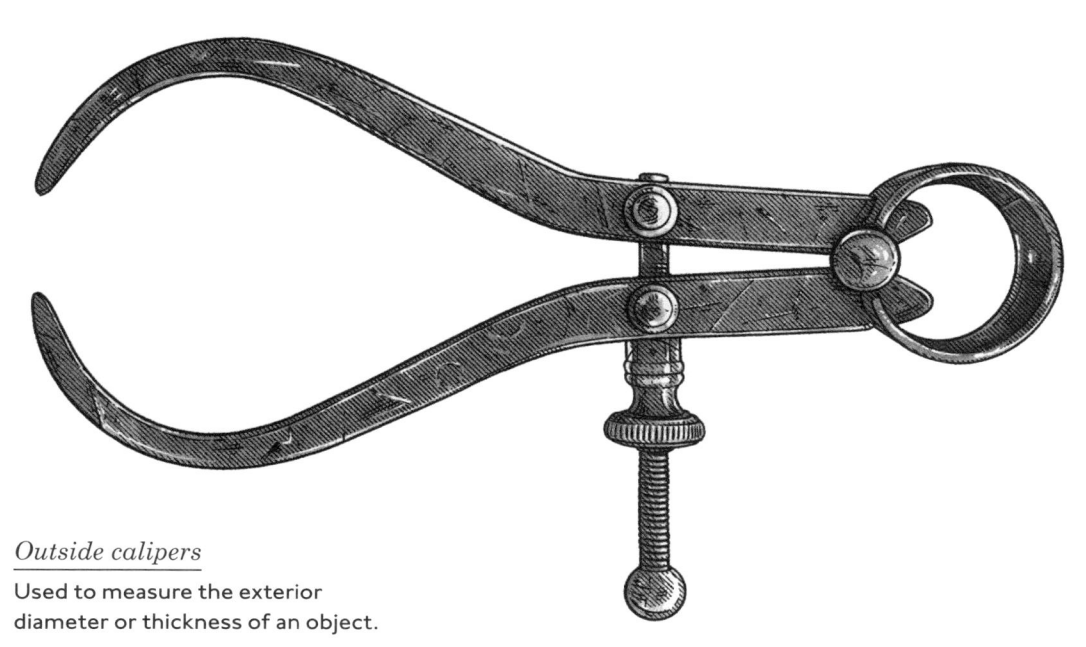

Outside calipers

Used to measure the exterior
diameter or thickness of an object.

Vernier calipers

A modern take on an ancient design,
the Vernier scale allows for very
prescise readings.

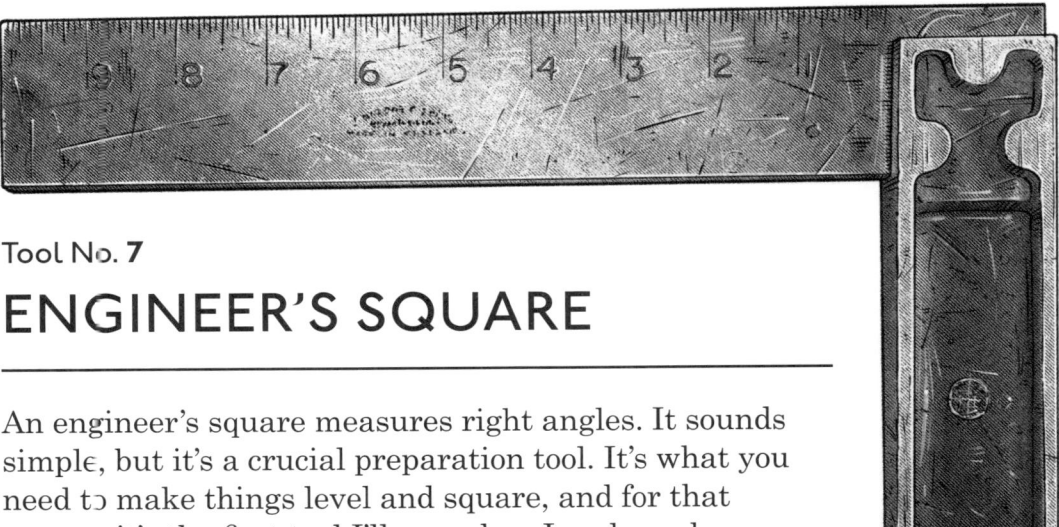

ENGINEER'S SQUARE

An engineer's square measures right angles. It sounds simple, but it's a crucial preparation tool. It's what you need to make things level and square, and for that reason, it's the first tool I'll use when I make or hang anything. The first mark you measure with it is the most important, because everything else will be orientated around it.

An engineer's square comprises two parts: the stock (base) and the blade, and they're usually made from stainless steel. They come in all sorts of different sizes depending on the size of the object that you're making or hanging, but the base is almost always quite heavy.

I buy engineer's squares all the time at car boot sales. Some are made of wood or brass and can look great when they're buffed up, but they're not as accurate as a solid steel one, which has a heavier base and is less likely to have been damaged. Stainless steel squares are also useful because they're magnetic and can deal with temperature extremes, which you're going to contend with if you're welding.

A problem with buying even a stainless steel carpenter's square second-hand is that you don't know if it's been dropped on the floor at some point, which can make it inaccurate, because you can't tell just by looking at it if the angle's off (unless it's properly battered). A trick to check that it's alright is to hold the square flat against a bench and draw a line with a pencil, then flip it over and do the same. If both of the lines line up, it's all good.

The largest one I've got has a base that measures about 60cm (2ft). I use it a lot because it's so solid you can clamp things to it, and you can trust it. And that's important, because you're often dealing with lots of unknowns, such as a new length of metal or timber, where you can't assume that anything is going to be straight. So it's great to have a tool you can rely on to tell you.

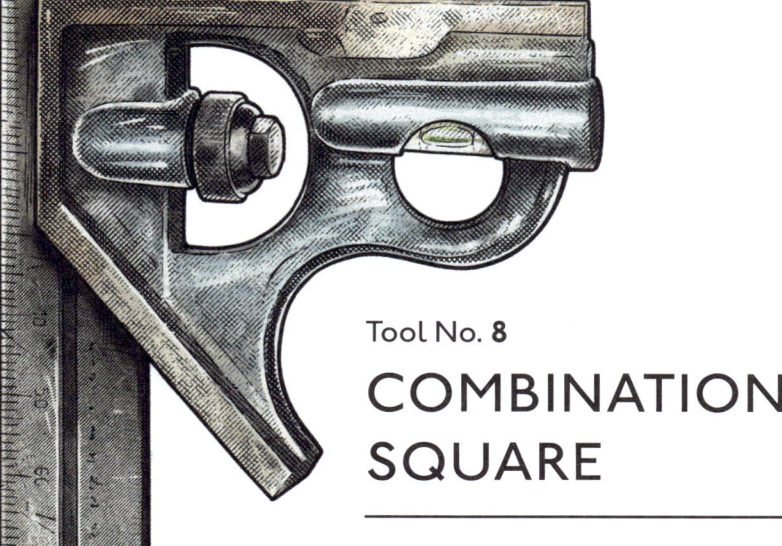

Tool No. **8**

COMBINATION SQUARE

A combination square does a bit more than an engineer's square (see opposite). It has a rule or blade, which is usually made of stainless steel, and interchangeable heads. The most common head is the standard head, which includes a 45° angle on a sloped part called the shoulder.

The 45° angle is useful when using the square for woodwork, for example, when you want to make a mitre joint by attaching two pieces of angled wood to form a corner. The standard head also features a spirit level to check whether or not a surface is horizontal or vertical, and an adjustable nut, which allows you to move the head along the length of the ruler depending on the size of the job.

The two other types of head are a protractor head, which is used for measuring angles, and a V-shaped centre-finder head, which helps you locate the centre of a circle.

The idea for the combination square was first thought up by Laroy S. Starrett in 1877, who set out to "invent something useful that people would want". He went on to found the L.S. Starrett Company in Massachusetts, which is still in business today. Looks like Starrett was right.

Tool No. **9**

CARPENTER'S PENCIL

A carpenter's pencil is basically a sturdier, more practical pencil than your classic classroom HB. The first thing you'll notice about it is that it's got flat sides, which stops it from rolling away and means there's a larger area to grip. It also fits behind your ear perfectly! Depending on how you hold it, you can draw either thick or thin lines, which makes it a versatile tool. And because of the way it's shaped, you can use it in narrower spaces than you can a regular pencil. It's also got a stronger core that won't break as easily, which is useful, because it's going to need to mark coarse surfaces and will probably get knocked about a bit by other tools.

You can get a specialized pencil sharpener for your carpenter's pencil, but you can also use a knife or whittling tool. One tip a lot of folks follow is to sharpen both ends of the pencil – one with a finer point and the other with a wider point, so you can switch between them easily and not have to sharpen from scratch if the pencil breaks.

As for the core of the pencil, it's still called "lead", which is confusing because but it isn't actually lead, it's graphite, which is a crystalline form of carbon. Part of the reason for the confusion is that when the first large quantity of solid graphite was found, in Borrowdale in the Lake District in 1565, it was thought to be lead. It was named *plumbago* (Latin for "lead ore") and translated into other languages, including Arabic, Gaelic and German. Their words for "pencil" still translate as "lead pen", so the lead/graphite confusion isn't unique to Britain!

The first pencil that looked like a modern carpenter's pencil is believed to have been made in 1560 by the Italian couple Simonio and Lyndiana Bernacotti. They cleverly hollowed out a stick of juniper wood and fitted a stick of graphite inside. The result was a thick, flat pencil that isn't a million miles away from what we use today.

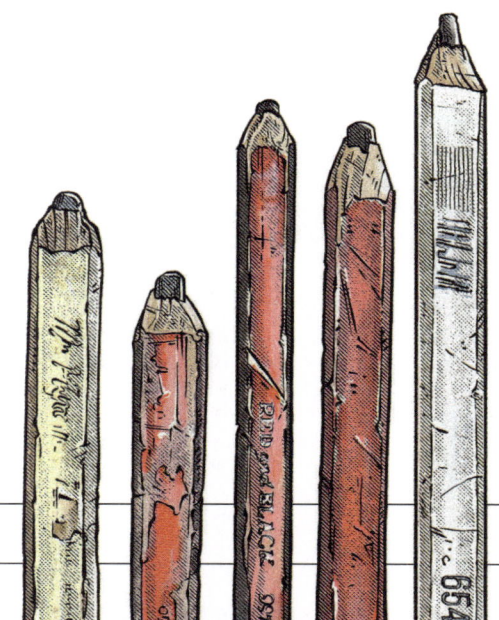

Tool No. 10

CHINAGRAPH PENCIL

These are known by lots of different names: grease pencils, wax pencils and china markers, but chinagraph is the one we tend to use in the UK. In the basic form that's been around for decades, it's a type of pencil made of hardened wax (usually paraffin or beeswax) surrounded by layers of paper, which can be unwrapped thanks to the handy string that dangles down so you can sharpen it.

The great thing about chinagraph pencils is that you can use them on all sorts of smooth and shiny surfaces, such as plastic, metal, porcelain and ceramics, and this is where the name comes from (they don't have anything to do with China the country). They are used by so many different types of craftspeople, from potters to glassblowers, who use these kinds of surfaces and want to make marks that are moisture-resistant. They're also useful in medicine for marking things such as x-rays and other scans that are often printed on glossy paper. Action films set on ships often feature your classic glass screen with the positions of ships and aircraft marked on with, you've guessed it, a chinagraph pencil.

In my early 20s, I was working at Karmann Konnection, a supplier of restoration parts and accessories for Volkswagen. When I'd take a headlight out of a classic 1960s Campervan, I could often see the chinagraph pencil marks the VW factory foreman had made years before to indicate which colour they should paint it. I loved that connection to the past, and since then, I've always had a fondness for chinagraph scribbles.

The pull-string type of chinagraph is cheap to buy but can become damaged easily, especially when you're rolling back the paper layers. These days, though, you can also get yourself a retractable grease pencil (sometimes called a mechanical grease pencil), where you push the top to expose more of the lead as and when you need it. This means you can refill it, too. You can also get a type of grease pencil that looks much more like a regular pencil, with the waxy core encased in a wooden tube.

Tool No. 11

TAPE MEASURE

Unlike a lot of tools, where their origins are tricky to uncover, we know quite a lot about the tape measure. It starts with a metalworker called James Chesterman, who moved from London to Sheffield in 1820 and later set up a factory that he called Bow Works, next to the river Don. We know that he had an affection for the bow (as in a bow and arrow), because his trademark included one encircling his initials.

Sometime in the mid 1820s, Chesterman came up with a new type of measuring device, and in 1829 he was granted a patent for his flat steel tape measure with measurements marked on to it. It wound up inside a round leather case, so it was easy to move around. He developed his design and began exporting it to the USA, but they were expensive, at $17 a pop (around 20 times that in today's money).

It was an American who went a step further. In 1868, Alvin J. Fellows of Connecticut patented a tape measure that had a spring that locked it in position and a button that released the spring. And so the tape measure, which works very much the same as modern boxy-style tape measures, was born.

I've got four old round leather tape measures with brass folding arms (see illustrations, opposite), three of which were made in Sheffield by James Chesterman & Co. One of these is a linen tape, measuring 100ft (30m), and the others are steel, one measuring 66ft (20m) and the other 50ft (15m). The fourth tape measure is linen, dates back to about 1940 and was made by John Rabone & Sons in Birmingham, who were most famous for their rulers (see page 10).

John Rabone & Sons supplied tape measures for the British war effort – if you ever come across one you can tell because they feature a small arrow on the brass handle, a symbol known as the "broad arrow" mark.

The original tape measures were made of flat steel wound up inside a leather case.

Tool No. 12

SPIRIT LEVEL

The spirit level – a sealed glass tube containing alcohol and an air bubble – was invented around 1661 by Frenchman Melchisédech Thévenot. He was a colourful character, serving as Ambassador to Rome in the 1650s and appointed the Royal Librarian to King Louis XIV in 1684. He also wrote a famous book on swimming that popularized breaststroke.

As with quite a few tools, it was the ancient Egyptians who got there first. They had come up with a very clever tool to help them measure level. It was an A-frame – three pieces of wood joined together that formed what we know as the letter "A" – and at the top of the "A" was a string attached to a weight that hung down below the horizontal piece of wood. If the surface the A-frame was resting on was level, the string would hang straight down across the middle of the horizontal piece of wood, which probably had a notch etched on to make it even clearer. And you can tell they got it right if you take a look at the Great Pyramid of Giza.

The Romans also used the A-frame design but went a step further. They built water tanks to use as spirit levels when they were constructing aqueducts, because they knew that the surface of undisturbed water is flat.

The A-frame was used all the way through to the 19th century in Europe. At some point, lead became the metal of choice for the weight suspended from the apex of the "A", and this is where the word "plumb", meaning exactly vertical, originates, sometime in the early 14th century. "Plumb" comes from the Old French word *plombe*, which comes from the Latin word *plumbum*, meaning lead.

Although the spirit level had been invented in 1661, it was mainly used for telescopes and specialized surveying equipment, and it wasn't until the Industrial Revolution that spirit levels produced in factories, resembling those we still use today, started to be used widely by carpenters.

Tool No. 13

THREAD PITCH GAUGE

Imagine I'm taking something apart that your grandad had made, to restore it. It's full of old nuts and bolts that he found in the shed 80 years ago, but in order to fit the holes I need to use the same-diameter bolts. This is where a thread pitch gauge comes in handy. It looks a bit like a Swiss Army knife, with two ends containing a number of different "leaves" with evenly spaced teeth. All you do is extend the leaf and see if the teeth fit the threaded part of the bolt. If they do, you're in business.

And it's so satisfying when you know exactly what nut or bolt your grandad was working with. Not least because I've got jam jars full of nuts and bolts on two shelves in the barn that have been waiting for this moment. And I think to myself, it was worth it getting up at 5am for that boot sale where I picked up that box of old bolts.

I prefer working that way, trying to keep something as true to its original form as possible rather than modifying it and making it modern. I love that lightbulb moment when something comes in on *The Repair Shop* and I know what I'm going to need to fix it and exactly where it is. I also love the link to the past, because I'll pick up a box of bolts, and I'll know from the type of bolt or even the thread what the person who owned the box in the past did professionally. Something like dome-head screws, which you can't buy anymore, would have been used by a woodworker. It's as if I've got snapshots of different people's sheds over a period of history.

Tool No. 14

CENTRE PUNCH

You use centre punches to make little holes or indents in a surface so the tip of your drill bit is nicely guided to where it's supposed to be. Centre punches are massively underrated preparation tools – they often come in a set with an electric drill or screwdriver, and often end up gathering dust at the bottom of the box. But these tools deserve a good home, too!

A centre punch is just a short steel bar with a pointed end. It has a "knurled" surface, which is a fancy way of saying it's got a pattern of little ridges on it so your finger can grip on to it. They've also usually got a square head, so they don't roll off your workbench. Whoever thought of that one is a genius.

I've got quite a few automatic centre punches that are spring-loaded, so when

you push down they recoil. You can adjust how much they "hammer" by twiddling the knob at the end. But most centre punches are simpler, and you just either press the end to where you need it or tap them gently with a hammer.

Nail punches look very similar to centre punches but their heads are flat, to match the head of a nail. They're used with a hammer to punch nails through wood.

Centre punches have a "knurled" or ridged surface, for easy gripping.

Tool No. 15

SCRATCH AWL

A scratch awl is a preparation tool used to scribe a line or shallow groove onto wood, leather and even sheet metal. It's basically a spike made of steel, with a sharpened point, and is used like you would hold a pencil.

After you've marked your line, you then cut along it with a saw, chisel or knife. Scratch awls can also be used by leather-workers to trace around a template design, which they'll go around later with a cutting knife. The tip is also small and versatile enough to get under stitches if you've made a mistake and find yourself needing to unpick your stitching.

Most modern scratch awls are made with a plastic handle, but some of the older ones have beautiful wooden handles, made of boxwood, ash, beech or African blackwood.

Some of these earlier awls have a chuck, so you can remove the steel tip and replace it with a different type of awl. You can also get heavy-duty scratch awls, which feature an enlarged handle, so you can get a proper grip with your whole hand.

You'll need to keep the end of the awl sharp, which you can do with a leather sharpening strop and fine-grit sandpaper. Just be careful, though (especially after you've sharpened it), because a scratch awl will roll off your workbench easily if you're not paying attention.

Before they were made of plastic, scratch awls had beautiful wooden handles.

Tool No. 16

STITCHING AWL

This begins with a bit of a depressing story, but I promise it gets better! Louis Braille was a chap you may have heard of because he invented the Braille reading system for the blind. His father, Simon-Réne, was a leatherworker and made harnesses, bridles and saddles for horses.

One day, Simon-Réne's three-year-old son Louis was in his dad's workshop and picked up a stitching awl, trying to make holes in a piece of leather like he's seen his dad do. While pressing the awl down, it bounced off the leather and caught him in the eye. The eye was damaged beyond repair and, tragically, the injury affected his other eye too, causing young Louis to go completely blind by the age of five.

Louis was an intelligent, positive and creative child who ended up attending the National Institute for Blind Youth in Paris in 1819. The school's founder had come up with a way of helping the children read by embossing thick paper with the imprints of Latin letters, which they could feel with their fingers. But the three books he created were expensive to make, fragile and didn't help the children learn to actually write by themselves. So Louis set to work on his own system inspired by a new alphabet of raised dots impressed onto paper, which was created by French inventor Charles Barbier.

Louis came up with the first version of his own reading system for the blind aged just 15, using the tools that Barbier had used to create his: a slate (two pieces of wood attached together by a hinge) and a stitching awl. So the system that ended up completely changing the lives of the visually impaired was first made using a similar implement to the one that robbed Louis of his sight.

A stitching awl looks similar to a screwdriver, with a narrow, tapered shaft and a sharp (or slightly bent) point. The earliest stitching awls, which were made from bone, date back to the Stone Age, and later ones made from iron have been found in ancient Rome. They're used for making holes or enlarging ones you've already made in leather, so they're commonly used by shoe repairers and other leatherworkers. There's also a stitching awl (also called a reamer) in the multi-purpose pocket knife (later known as a Swiss Army knife) that was first issued to the army of Switzerland in the 1890s.

The earliest stitching awls, which were made from bone, date back to the Stone Age.

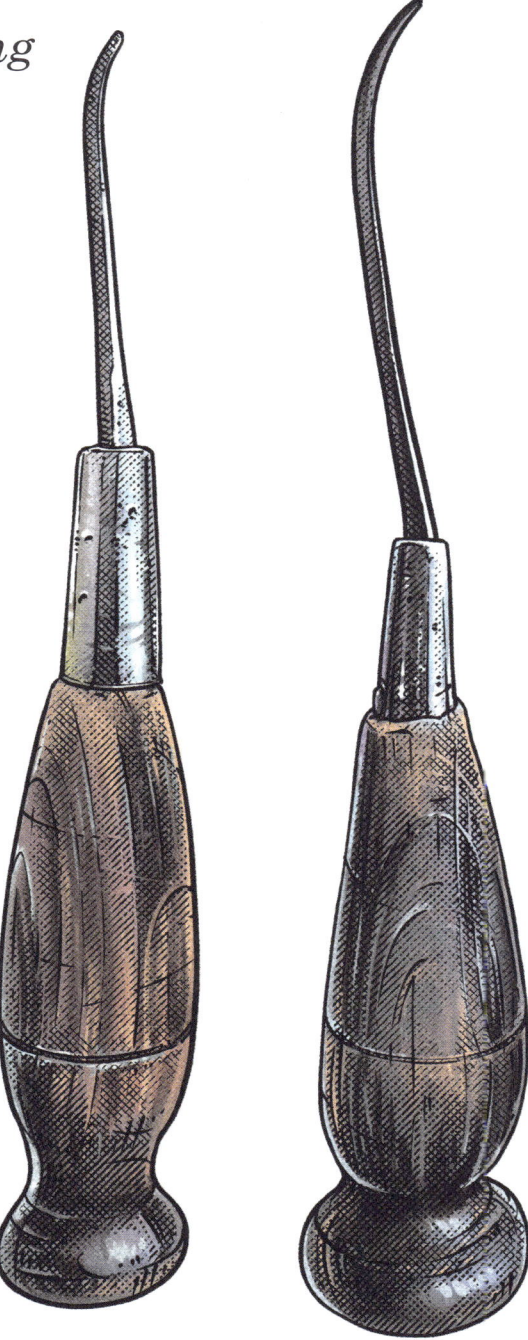

Sritching awls

A stitching awl has a narrow, tapered shaft with a sharp point, whereas a scratch awl (see page 25) has a straight shaft.

CHAPTER 2
GRIPPING & HOLDING

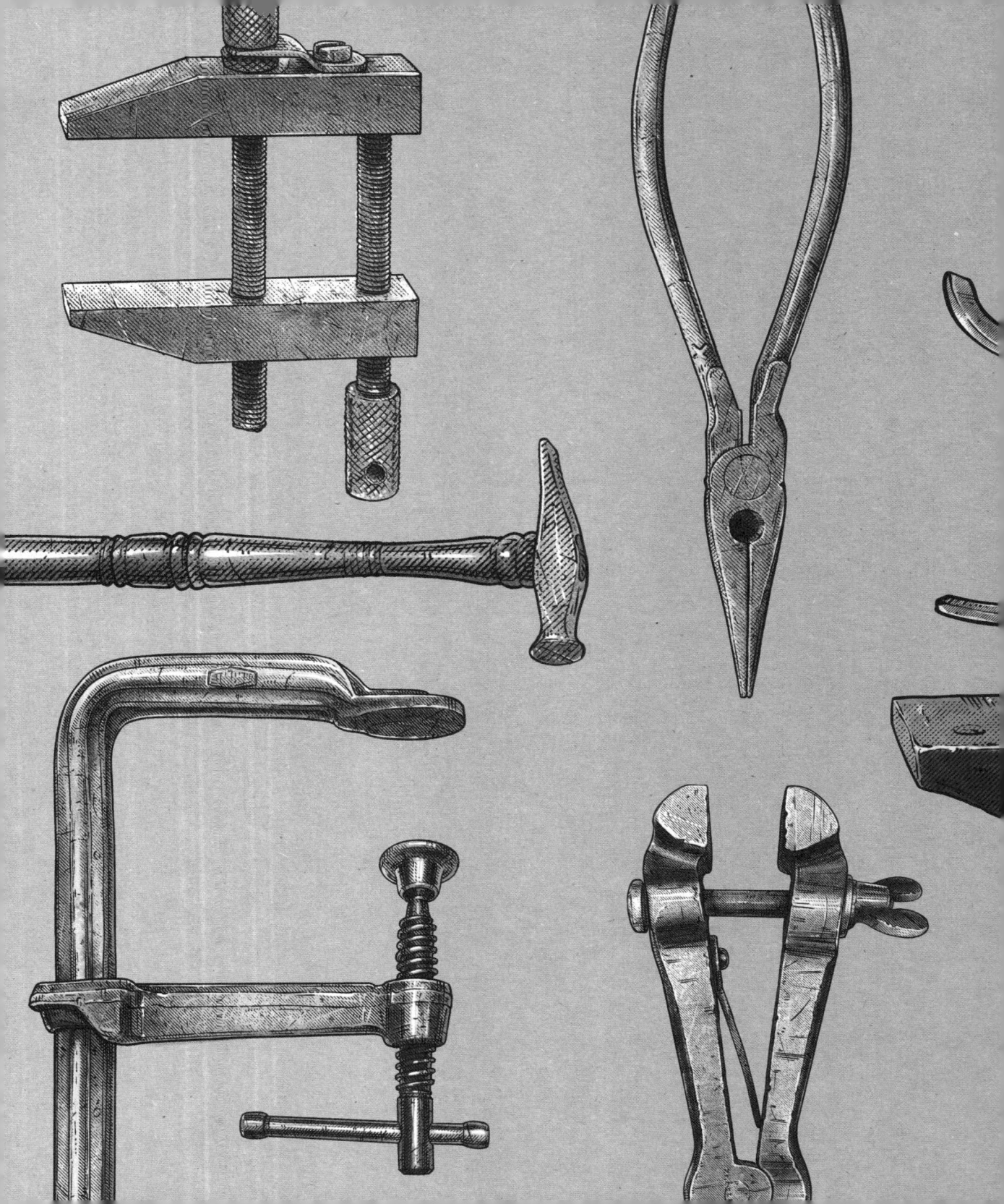

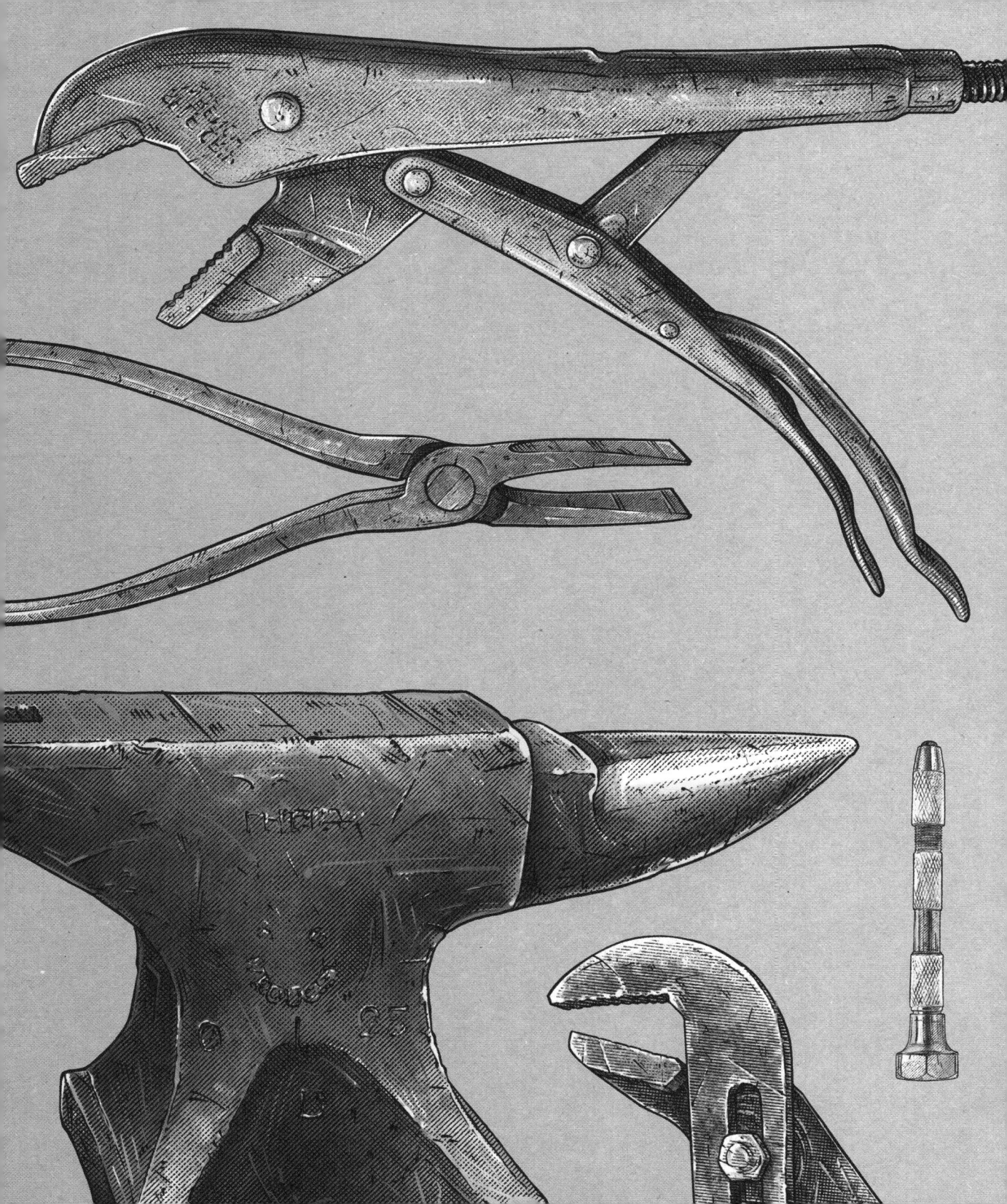

Tool No. 17

BENCH VICE

Bench vices mean a lot to me. This is my favourite vice – it's the one I use the most and the one I got from my dad. Also, it's got a set of tube-bending jaws under the main jaws, which often come in handy for jobs on *The Repair Shop*.

I got another bench vice from a guy I'd bought my first Ranalah English wheel from. The vice was in a sorry state, living in a muddy patch of ground in a big field that looked like a graveyard of broken, damaged and decaying tools. I remember he had a big old rusty Bridgeport mill that would have once been worth a fortune. The whole place felt like it was full of broken dreams. Everywhere you looked, inside and out, there were tools and half-started projects. Many years ago he'd begun rebuilding his house, but he'd never got anywhere near finishing it, and the scaffolding that had been up for years had gone rotten. The caravan that he'd moved in to temporarily had become his home. The whole experience stayed with me, and I took it as a warning not to get involved in too many projects at the same time.

I pointed to something that looked vaguely vice-like in the garden, but to be honest, it was so rusty it could have been anything. Upon closer inspection, everything had seized up and it wouldn't work at all, but he knew exactly where it had come from and what he'd used it for. I ended up buying it because I knew it could be a lovely vice again, and got to work on restoring it. We tried using electrolysis to free it up, but it wasn't having any of it, so we ended up heating it up and bashing it.

It turned out that underneath all the flakes of rust was a lovely, old, British-made bench vice with a quick-release lever. The company logo was completely invisible until we'd taken enough of the rust away; I won't forget that moment when you could see the Parkinson's mark come in to view – Parkinson's is a big name in vices because it was Joseph Parkinson, from Bradford, who invented the quick-release vice in 1884. Over the next couple of years, he patented his invention in the US and Canada, among other countries.

Getting the vice to the point it's at now has taken a lot of time and money, but it was massively worth it, and the vice is once again a beautiful tool. That whole process of salvage and restoration is always a unique journey, on which you begin to form a relationship with the tool. With a vice in particular, you have to trust that it won't give way at a crucial point, and over time, you get to know its quirks, like if it needs half a turn before the actual jaws start moving.

The vice I value the most is in the wagon shed (an open-fronted building used in the 18th century to house carts and wagons needed on a farm) at the Weald & Downland Living Museum, where we shoot some scenes for *The Repair Shop*. Inside the shed is my vice, attached to a large workbench. It used to live in my dad's shed, and it was made before I was born. When we've finished filming for a bit, it comes back with me to my workshop. I've got three other vices at *The Repair Shop*. I've got better vices – quick-release ones, larger ones, one with big anvil pieces on the back – but there's something about this vice that makes it the go-to one for me. It's probably because it was my dad's and from my family home that it means so much to me. If ever I've got a tricky situation and I have to rely on a vice, that's the one I choose. It's got me out of trouble so many times!

Underneath the main jaws is a smaller set of jaws that I use for bending tubes, which gives the vice extra usefulness. I remember needing to make a push-along handle that went up and over for a teddy bear on *The Repair Shop* and it was the perfect size. When something like

that happens, and the vice proves to be the perfect tool for the job I'm working on, I get more and more attached to it.

But vices don't just secure things. I use them to press bearings in and out, and sometimes I use them for folding metal if it only needs to be done crudely. Some bench vices have got anvil blocks at the back, but they're only suitable for light work when you need a flat surface to hammer something flat. And having an anvil block stops you from using the hollow casing that you'll often find at the back of the vice, which won't stand up to hammering, however tempting it is to use it when you're tired and can't be bothered to move it to a proper anvil or workbench!

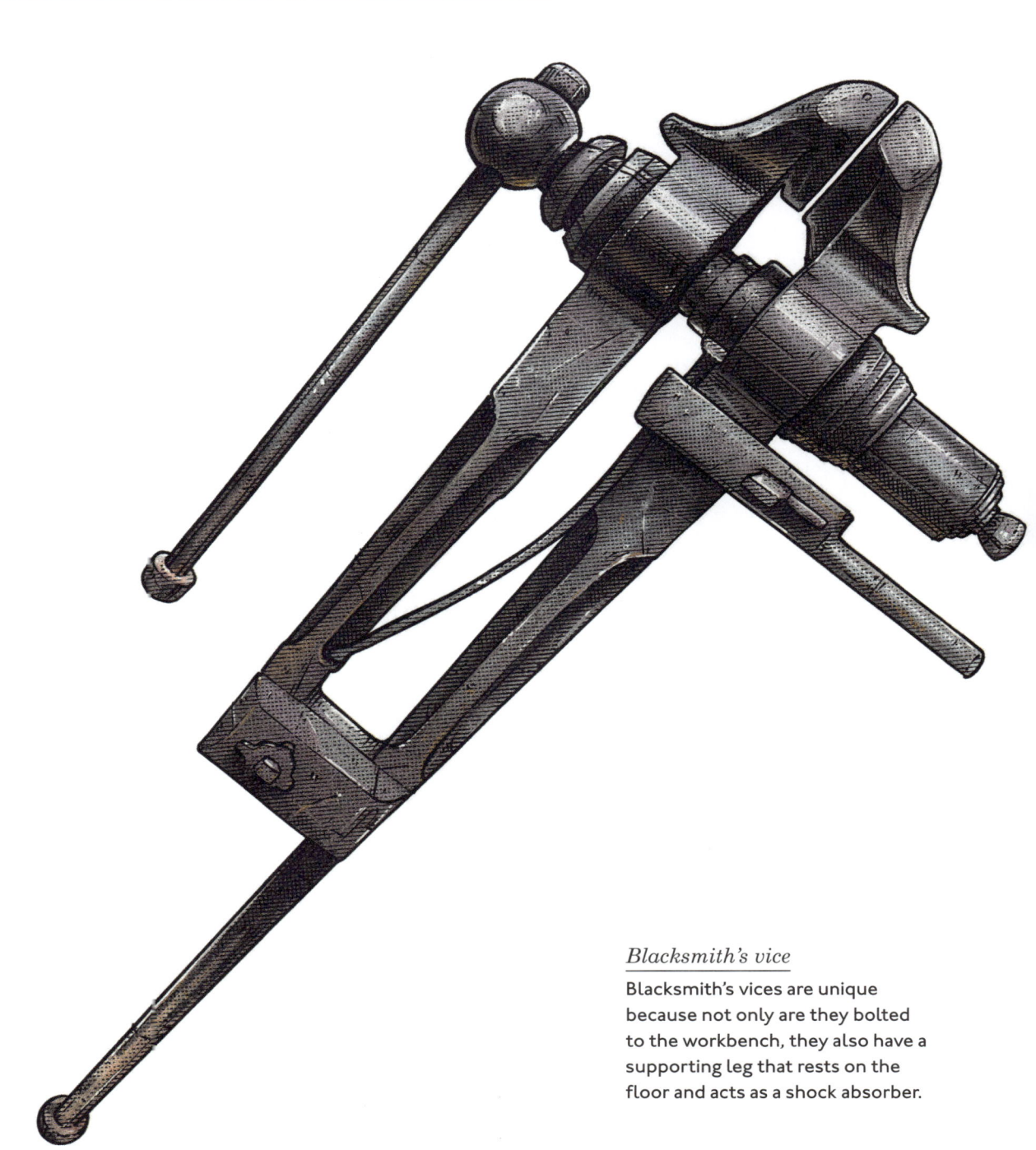

Blacksmith's vice

Blacksmith's vices are unique
because not only are they bolted
to the workbench, they also have a
supporting leg that rests on the
floor and acts as a shock absorber.

Tool No. 18

BLACKSMITH'S VICE

The other vice that I don't have but really would like is a blacksmith's vice – it's a properly heavy-duty bit of kit. Blacksmith's vices are unique because, unlike most other vices, they have a support leg that rests on the floor as well as being bolted to the bench. They're also typically made of forged steel rather than cast iron or ductile iron. Forged steel is an alloy of carbon and iron that is compressed under pressure, and that makes a big difference, because its tight grain structure means it is mechanically very strong.

A blacksmith's vice needs these enhancements because it takes more punishment that any other vice, being hammered on all day long. However, because they're so sturdy, with the support leg, the shock transfers through the leg to the floor, so you don't feel any movement or bounce. Although blacksmiths have been around since the Bronze Age, the vice (in the sense that we'd use it now, to describe a tool with sliding jaws that moved horizontally) wasn't invented until the 18th century, and the first vices were made of wood. The cast-iron vice followed in the first half of the 19th century. By 1837,

we know that in the USA, the company of Wallingford & Co. was advertising "Lamont's patent solid box vices", although it wasn't until the 1840s that the solid box vice as we'd know it today, with a threaded nut and flange, was developed. The credit for that goes to Peter Wright, an anvil and vice maker from Dudley, near Birmingham, who received a patent in 1863 for his solid (one piece) box. We'll explore Peter Wright in more detail when I talk about anvils on pages 45–49. Meanwhile, if a Peter Wright leg vice from 1863 turns up in a car boot sale, it's going straight in the van!

A blacksmith's vice takes more punishment than any other vice, being hammered all day long.

PIPE VICE

A pipe vice is designed to hold a pipe safely so you can either cut it or cut threads in to it, to provide a strong seal when connecting it to another pipe. They can also be used for welding pipes together. There are two types of pipe vice: hinged and chain. Both are really important tools for plumbers.

A hinged pipe vice secures a pipe between two V-shaped sets of ridged jaws, which are moved by a handle at the top. The vice is usually mounted on a workbench or table. A chain pipe vice does the same job but secures the pipe with a steel chain that loops around it. They tend to be used for thick pipes that won't fit inside a hinged pipe vice. They can also be fixed to a workbench but are often attached to a portable tripod stand.

One of the more famous makers of pipe vices is Record, a brand name that most people have seen at some point. The company that registered the Record trademark (in 1909), was C. and J. Hampton, established in 1898 in Sheffield. They specialized in making good-quality clamps, cramps and vices between 1909 and 1930, and after that began to diversify, producing lovely hand planes in 1931.

Closely related to a pipe vice is a V-block, which is usually a rectangular steel block with a V-shaped channel at the top so you can hold pipes or tubes in position. The V-shaped channel looks like the lower jaw of a pipe vice, and I use these all the time for metalwork. You can also get fancy V-blocks that have clamps on the top, or even ones with internal magnets, so you don't need to physically clamp a pipe or tube in position.

A hinged pipe vice secures a pipe between two V-shaped sets of ridged jaws, which are moved by a handle.

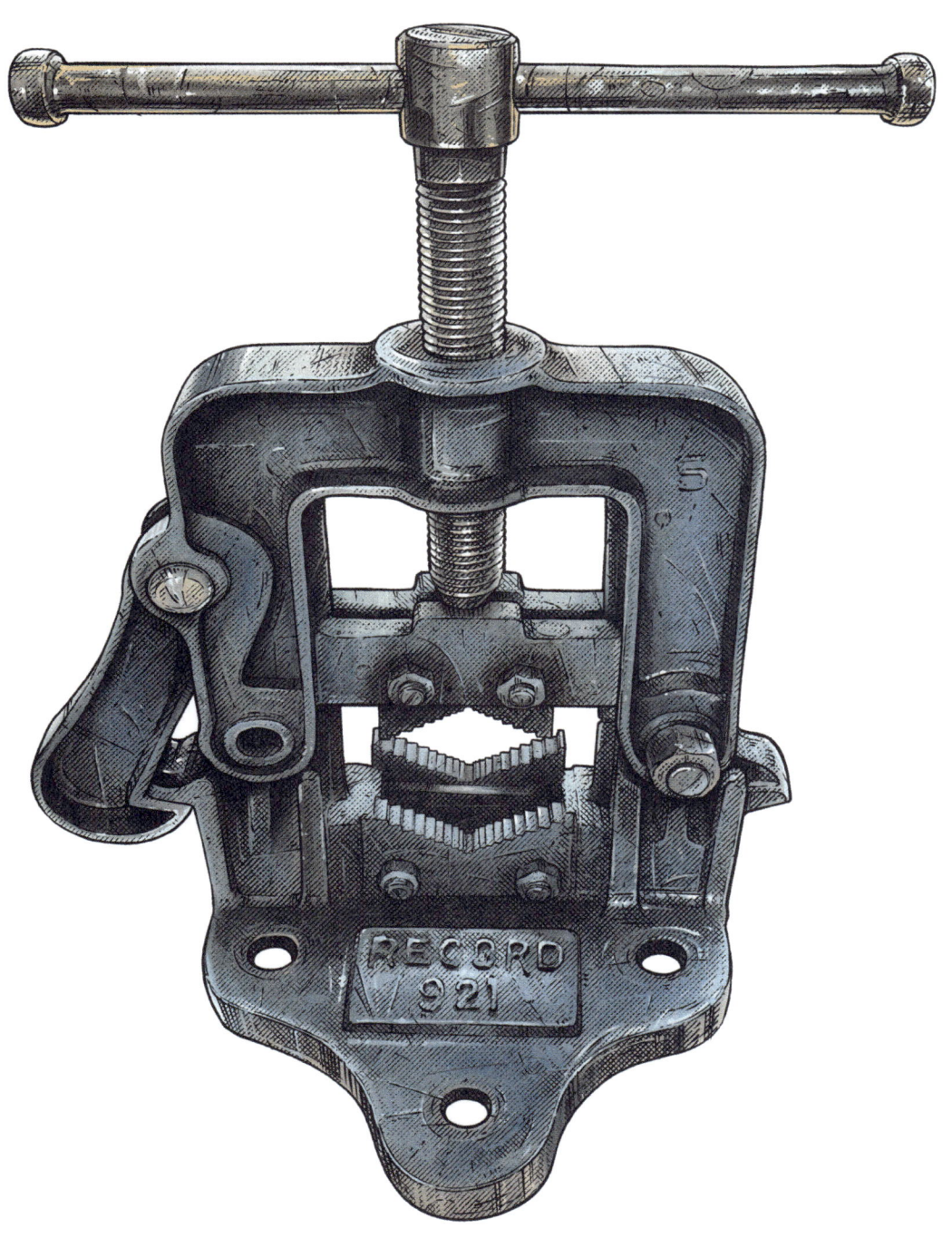

HAND VICE

A hand vice is a small, versatile tool that is designed to be used with one hand. It looks like a pair of tongs but with jaws instead of circular ends that are tightened and loosened with a wing nut.

Hand vices usually have ridges on the jaws so they can grip circular items easily. They are really useful when you're working on small, delicate items, or if you want a safety barrier between yourself and another tool you're using. For example, when you're using a grinding wheel to grind a small object, or if you're using a soldering iron in your other hand. In the past, they would have been used mainly by mechanics, engineers and jewellers.

One of the oldest hand vices I've been able to find out about is the Stevens hand vice, invented by William X. Stevens of Brookfield, Massachusetts, who received a US patent for the tool in 1869. It got a glowing write-up in the *Scientific American*, the oldest continuously published magazine in the US, only two months after the patent was granted. One of the reasons the writer loved it so much was because the handle "may be slipped off and the shank inserted in a half-inch hole in the bench, when it becomes a neat permanent vise for light work". There was a vertical groove in the face of the jaws that was perfect as a drill chuck, which served "a great long need felt by mechanics and amateurs".

In the UK, the company Thewlis & Griffith of Warrington was manufacturing

Hand vice

Unlike most vices, the small handheld vice is designed for close, delicate work and for gripping objects that are too small to be held by a bench vice.

tools from 1835 and started producing beautiful spring-loaded cast steel hand vices sometime in the late 19th century. Some feature the broad arrow symbol, so we know they would have been bought by the War Office and used in any number of the conflicts Britain fought during that period.

Another company around at that time was Ward & Payne of Hillsborough, Sheffield, which also made forged hand tools from 1843. Ward & Payne was sold to Wilkinson Sword Ltd in 1967, which continued to manufacture tools on the same site until a fire destroyed the works in 1970.

Tool No. **21**

PIN VICE

A pin vice is really useful when you're working with jewellery, watches, models or miniatures. Steve Fletcher, the resident clock restorer at *The Repair Shop,* uses them a lot.

A pin vice is basically used to lock drill bits in place, turning it into a small manual drill that can be used to drill very precise holes in very small objects. But it can also hold other very small tools, such as files or reamers (used to enlarge holes that have been drilled), or the actual object you're working on.

There are several different types of pin vice, but the most common is the swivel-head pin vice. The tip of this pin vice is a screw-on head that covers an adjustable chuck. You unscrew the head, which loosens the chuck, and then insert the drill bit or other tool. To use it, you position the swivel head in your palm, holding the barrel of the vice with your fingers and turning it with your thumb and index finger.

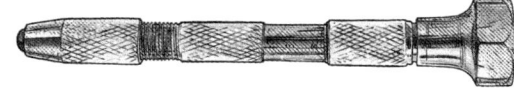

The most common type of pin vice is the swivel-head, which has a srew-on head covering an adjustable chuck.

Tool No. **22**

MACHINE VICE

This type of vice is used to hold objects securely when you're using a machine tool such as a grinder, milling machine or a drill press.

....

Machine vices are mounted on the surface of the actual machine tool rather than a workbench and feature hardened jaws to deal with the pressure they're going to be under. You can fit jaw pads, which are made of plastic, over the jaws so they don't damage the surface if you're working with something delicate. Some machine vices have a tilting jaw, so you can securely hold objects that are oddly shaped. Other machine vices even have a swivel base so the whole vice can be set at any angle.

Machine vice

You'll find machine vices in every variation, from simple single-jaw flat vices designed to hold one object at a time, to fancy multi-jaw devices for bigger jobs.

F-clamp
F-clamps were invented by Bessey, the German tool manufacturer founded by Max Bessey, in 1936.

F-CLAMP

This classic tool takes me back to woodworking lessons at school. An F-clamp is similar to a G-clamp (see page 42) but has a wider opening, so you can use it for larger objects.

It's called an F-clamp because of the shape it forms when the jaws are open, but it's also known as a bar clamp because it's made of three separate bars – one vertical and two horizontal. The higher horizontal bar is fixed, while the lower one is adjustable and attaches to a screw that tightens the clamp.

F-clamps get used a lot in woodworking, and the clamping plates on the jaws are often fixed with jaw pads to protect the surface of the wood. Some F-clamps have a tilting plate on the adjustable jaw, so you can clamp onto a tilted surface, and some feature a spring-lock trigger, which you press to make the movable jaw slide quickly along the frame. You release the trigger to lock the jaw back into the position you want it in.

Tool No. **24**

G-CLAMP

I love this tool, and it's a classic for a reason: it's simple, strong and the old steel and cast-iron ones are a thing of beauty. They take you back either to your first design technology class or your grandad's old tool shed, where they'd be hanging from the rafters.

It's called a C- or G-clamp because of its shape. There's usually a flat surface at the top right of the "C", and at the bottom right is a threaded hole. A screw, with the bottom end fixed to a metal bar and the top end forming a flat edge, goes through the hole. You turn the screw to the right to tighten it. When the screw goes up through the threaded hole, the shape of the clamp looks like a "G".

They're used for all sorts of things, but are probably most associated with holding pieces of wood and metal in place. If you are using them with wood, be careful to place offcuts of wood between the clamp and the wood you're working on so the surface doesn't get marked. I use G-clamps all the time when I'm working with metal. However good you are at welding, a G-clamp is vital because everything you're working with gets too hot for you to hold yourself, so you come to rely on a good old G-clamp (or ten).

The older, sturdier, heavier G-clamps were usually made of high-quality steel and were built to last by folk who took a lot of pride in their work. And that's why they're doing exactly the same jobs just as well as they were 50 years ago. The newer, cheaper ones are often made of lower-quality metal and they can sometimes buckle at the top left of the "G".

Quite often, when I'm making something from scratch, I'll do a dry build, which means I assemble the components of what I'm making first but without using any screws or any glue. I'll clamp everything together using G-clamps to make sure it's all lining up, everything's the size it's supposed to be and fits where it's supposed to fit. Then if there's anything wrong, I can just unclamp it and make the adjustments I need to.

G-clamps are used a lot in film, TV and on stage to attach heavy lighting fixtures to long metal pipes, called battens. Back in my former life as a set designer, I'd use them everywhere.

G-clamps are used a lot in film, TV and on stage to hold lighting fixtures in place.

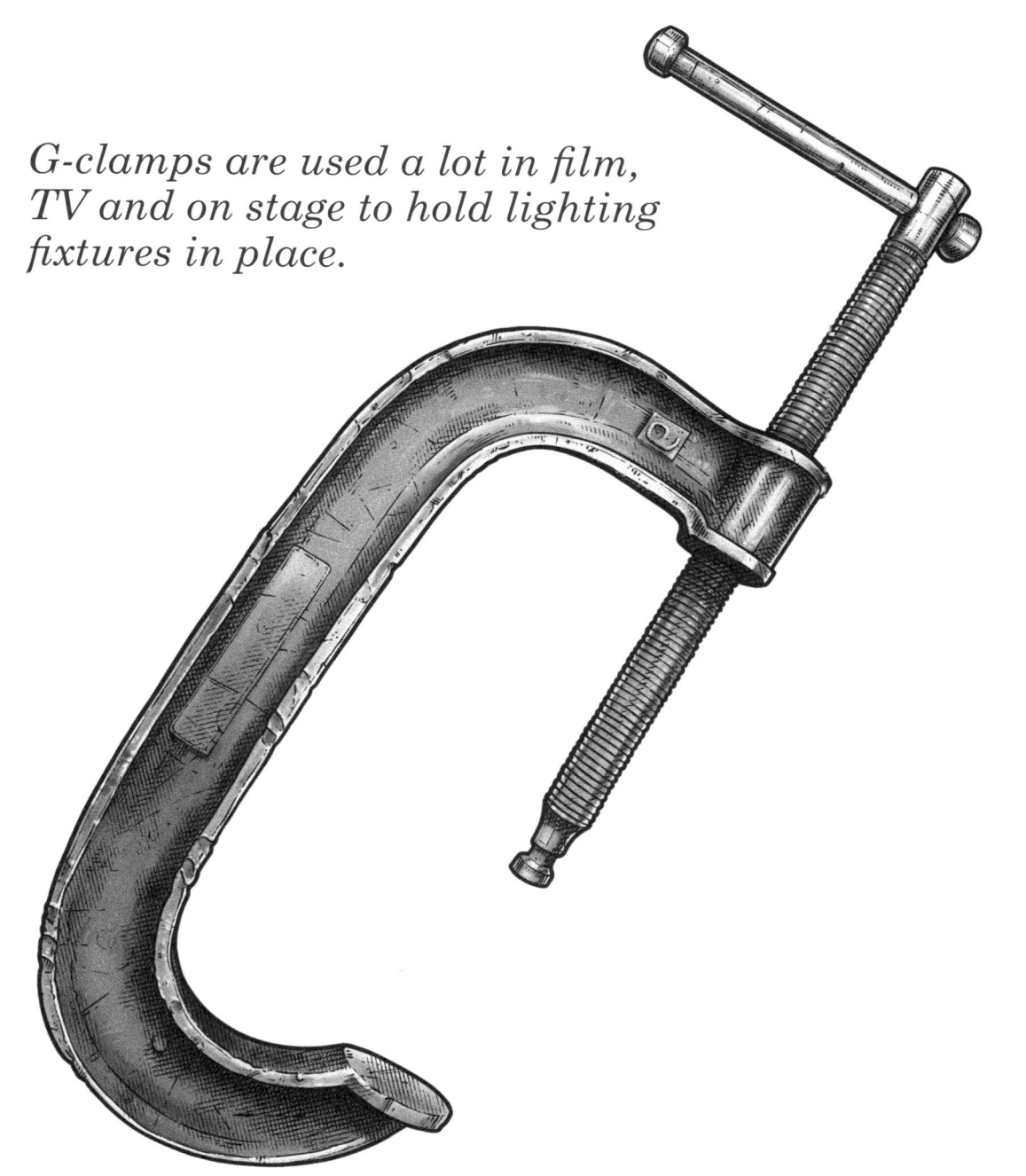

TOOLMAKER'S CLAMP

These are usually slim, strong clamps that are most commonly used by (you've got it) toolmakers, who often work with little components. Having said that, you can get absolutely massive toolmaker's clamps, with huge jaws – it depends on the type of tool you're making.

I adapted a toolmaker's clamp recently in my little barn at the set of *The Repair Shop* by cutting the ends and welding a piece of metal on to get a pin out of something (it was the only tool that could do it) and I've used it a few times since. They're brilliant because they've got parallel jaws and are great for very precise jobs. These versatile little clamps sit nice and flat and square, and there's no big handle sticking out the side. They're actually one of the tools that you learn to make quite early on when you learn toolmaking, because they're so simple. They've got two screws – one in the centre and an outer one, which you turn to change the width between the two jaws.

Toolmaker's clamps are used for much more than toolmaking though. Engineers use them, and Steve, our resident clock specialist at *The Repair Shop*, has a whole bunch of them he uses for various jobs.

These handy little clamps are great for very precise work and sit nice and flat on the work surface.

JEWELLER'S ANVIL

A jeweller's anvil is much smaller and lighter than a blacksmith's anvil (see page 46). Some tiny ones weigh only a few ounces, but most are around the 1.25kg (2½lb) to 5kg (10lb) mark. Although they're small, they're made from forged steel or cast steel, so they're very strong.

Shape-wise, jeweller's anvils usually have a squarish body, and a long, thin heel and horn. The flat surface is finely polished, so it's as smooth as possible to work and shape delicate jewellery. Some jeweller's anvils have a large base, but many are designed with a stake at the bottom so they can be embedded into a hardwood stump or a bench.

Although they're called jeweller's anvils, they're used by quite a number of different crafts, so they're sometimes known as silversmith's, clockmaker's and watchmaker's anvils. While their shape hasn't changed very much over the past 300 years, some more unusually designed ones have been produced. John Arnold, the famous watchmaker, was using mini anvils with a circular body, three horns and a heel in his workshop off London's Strand in the second half of the 18th century. They would have most likely been used to manufacture his marine chronometers, the third of which went with Captain Cook on his second voyage to the southern Pacific Ocean in 1772–7. Very occasionally, these anvils, marked with F. Arnold & Sons, come up for sale, but they aren't cheap!

BLACKSMITH'S ANVIL

Everyone knows what a blacksmith's anvil looks like, with its flat face at the front and horn at the back. Maybe that's got something to do with anvils appearing as a joke in Warner Bros cartoons as a ridiculously heavy object. There's one cartoon I remember of Sylvester the Cat seeing Tweetie Pie in the window of a skyscraper, so Sylvester does what any hungry cat would do and starts blowing bubblegum into a bubble so big that it makes him float upwards. When he reaches the window, Tweetie Pie helpfully hands him an anvil, and he plummets. Classic.

An anvil is basically a blacksmith's work table, and it has to be really strong to stand up to the constant hammering that it's going to have deal with. It's weight and size varies a lot depending on what it is going to be used for. We're talking 9kg (20lbs) all the way up to 450kg (1,000lbs), but most blacksmith's anvils weigh between 45kg (100lbs) and 135kg (300 lbs).

Modern anvils are made from cast steel, which involves heating steel until it's molten, then pouring it into a mould. Now that all sounds easy enough, but it took a long time for us humans to get to this point, so let's rewind a bit.

Early blacksmiths would have used stone as a striking surface, followed by bronze sometime after 3,200 BCE (three dozen anvils have been found in Europe dating from 1,200 to 700 BCE). Blacksmiths would have used bronze anvils to manufacture small bronze and gold items such as jewellery and weapons.

Meanwhile, the Hittites in Anatolia (now Turkey) learned how to smelt iron around 1,500 BCE, but this didn't exactly happen overnight – it took some time to learn how to work with iron because early furnaces weren't hot enough to melt it. At first, they just ended up with something called bloom – a spongy mixture of iron and slag – which they couldn't do much with. Eventually, heating and repeated hammering got rid of the slag and they ended up with wrought iron. This wasn't an instant triumph, because it turned out that it was softer than bronze, you couldn't sharpen it and it rusted quickly.

More trial and error followed until some clever blacksmiths discovered that cooling hot iron in cold water made it less brittle. Later, blacksmiths working near the Black Sea discovered that inserting iron bars into white-hot charcoal created something much stronger than anything they had worked with so far: steel! With this massive leap, anvils started to be made with steel faces, and these were produced up until the 20th century, when cast steel became the material of choice.

The classic shape of the anvil – with its "horn" at one end, flat surface at the other and curved body tapering down to its solid base – became known as the "London Pattern" in the 19th century. But the rough shape had been around since at least the 16th century.

There are two holes in the face of an anvil. The square one is called the hardy hole, which is where you place square-bottomed, specialized cutting tools (called hardy tools) such as chisels. The other, round hole is called the pritchel hole and is designed so you can use a punching tool called a pritchel directly over it. You can also use the hole to rest tools with round bottoms. The distinctive horn of the anvil allows a blacksmith to hammer curves into a metal object, and the raised flat surface between the horn and the face of the anvil, called the "step" or "table", is often softer and is used as a cutting area.

In 1830 Peter Wright (see page 35) came up with the "solid anvil", made of one piece of forged iron. Most of Peter Wright's anvils weighed between 36kg (80lb) and 68kg (150lbs) so they could be moved about if needed. They're beautiful tools and go for a lot of money on vintage tool sites.

Tool No. **28**

FARRIER'S ANVIL

A farrier is a craftsperson who looks after horses' hooves and fits them with shoes if needed. They need similar skills to a blacksmith, but also have to learn a lot about the anatomy of horses' limbs. And like a blacksmith, they need an anvil, which they use to forge horseshoes.

A farrier's anvil is lighter than a blacksmith's anvil (see page 46) and usually weighs about 30–45kg (70–100lbs) so it can be moved about. That suits most modern farriers, because they tend to travel to the horse rather than the other way round. But a farrier's anvil isn't just smaller than a blacksmith's anvil, it's also highly specialized to suit farriers.

Firstly, it usually has a much narrower waist – the part where the base connects to the face (or horn) than a blacksmith's anvil. Also, the heel (the far edge of the face) of a farrier's anvil is usually longer and thinner, and its horn is more swollen-looking than a blacksmith's anvil, so most of the weight is in the horn. The face may also contain two pritchel holes (rather than the one you find on a blacksmith's anvil), which are used to make nail holes for horseshoes. Modern farrier's anvils often have two "turning cams" (small circular protrusions) on the body beneath the face, which look a bit like big, round buttons. These are used for adjusting horseshoes. The table (or step) of the anvil might also feature a clip horn for forming toe clips.

It's worth remembering that in the UK, it's actually illegal to call yourself a farrier or carry out any work that a farrier might do unless you've had four years' training, served an apprenticeship with an established farrier and registered with the Farriers Registration Council.

A farrier's anvil is highly specialized to suit the work of fitting horseshoes.

A farrier's anvil may look similar to a
blacksmith's anvil (see page 46), but a
crucial difference is that it's much lighter
so it can be moved around easily.

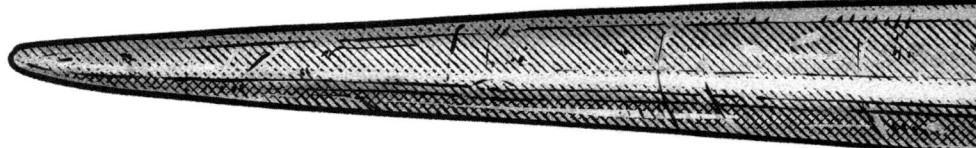

Tool No. **29**

SHEET-METAL ANVIL STAKE

These anvils, designed for working with sheet metal and doing decorative iron work, have a point rather than a base. That's because they're made to be embedded in a bench, supported in a stake plate or mounted in a wooden stump. They can be held in a strong vice, though, if the work you're carrying out is quite light.

These anvils vary a lot in shape, but one of the more familiar is the blowhorn stake, so called because it looks like something a Viking might blow to send the fear of the gods into the enemy. But they're actually used to make brass musical instruments, such as French horns and euphoniums. The blowhorn stake also has a very long tapered heel that is designed to form rings and other hollow, round objects. The face is very small and is typically used for riveting.

The beakhorn stake anvil looks similar to the blowhorn, but the horn is much flatter and narrower and the heel isn't as long. They're used to fashion metal into a rounded shape, such as curved handles or the cylindrical part of a drinking mug.

The funnel stake is another anvil worth mentioning. Despite the name, it isn't actually shaped like a funnel, but is used for forming conical shapes like funnels from sheet metal. Half-moon stakes are the anvil of choice if you need to curve the edges for sheet metal, and, in contrast, round bottom stakes are used for squaring up edges.

Some stake anvils have been discovered that date back to the Bronze Age. A few have been found with two points so the anvil itself can be used in two positions. Cleverly, the metal used to make the stake was shaped and sometimes ridged so that it provided a second work surface when the other point was in use. So, if you think a multipurpose tool is a modern invention, you're wrong!

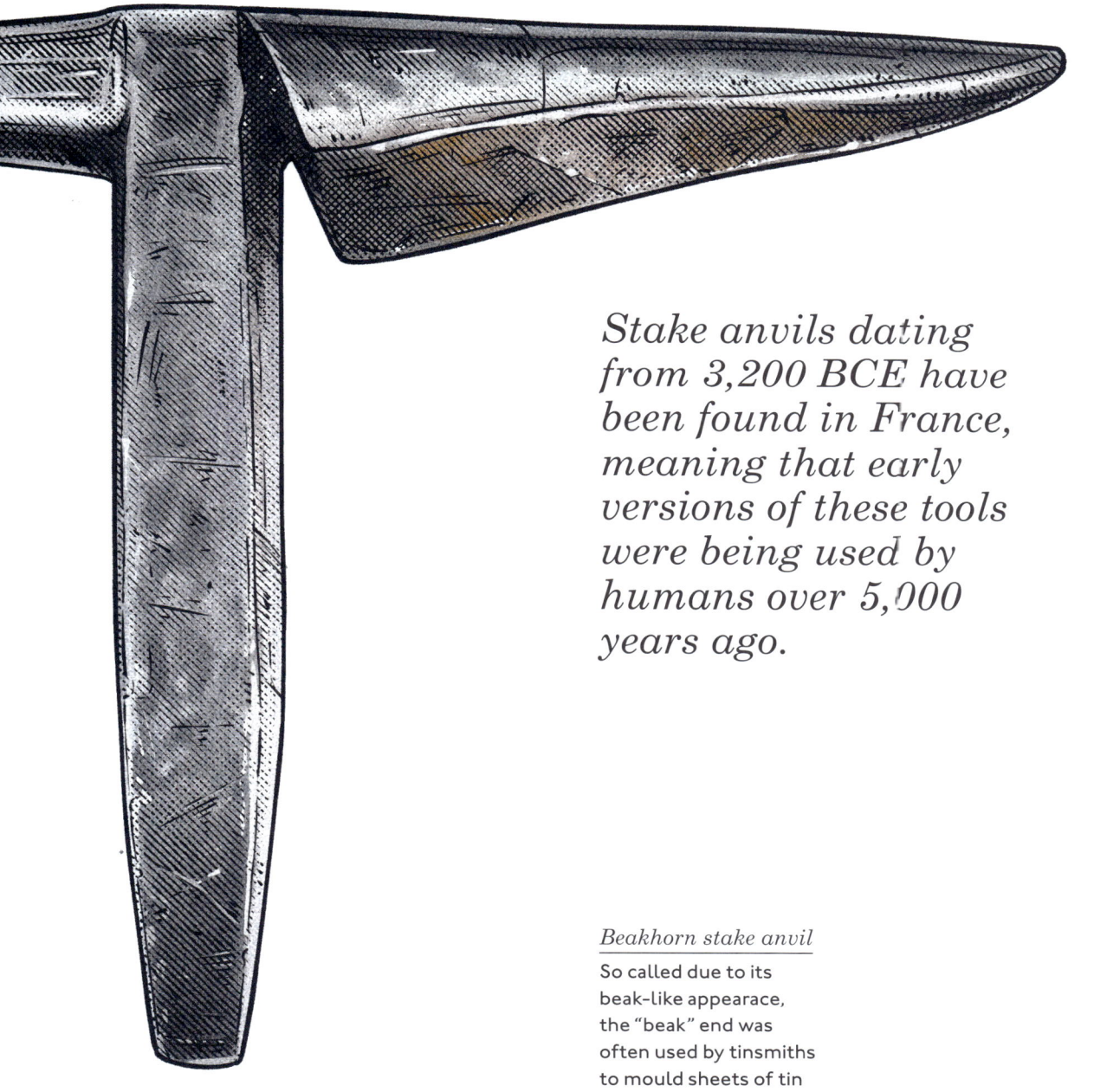

Stake anvils dating from 3,200 BCE have been found in France, meaning that early versions of these tools were being used by humans over 5,000 years ago.

Beakhorn stake anvil

So called due to its beak-like appearace, the "beak" end was often used by tinsmiths to mould sheets of tin into rounded shapes.

MOLE GRIP

A mole grip is basically a strong pair of pliers, which are usually made of hardened steel, and can be locked into place to hold something still. The toothed jaws are too aggressive for woodwork and will mark wood, but they're perfect for metalwork. Mole grips are especially useful when you're drilling, because they will lock whatever you're drilling in place, rather than having it spinning around in time with the drill. There's nothing more annoying!

I use mole grips a lot when I'm welding, because the heat from the welder can twist and distort the metal I'm working on. So I put as many mole grips as I can in place to minimize the amount the metal is going to move.

As for the history of the mole grip, you'd be forgiven for thinking that it was invented by a Mr Mole. But it wasn't. Instead, the mole grip was patented in 1955 by a Scottish chap called Thomas Robb Coughtrie who lived from 1917 to 2008 and was an electrical engineer for the Birmingham-based engineering company M.K. Mole & Son. Coughtrie became the managing director in 1950 after the second of the two Mole brothers died. It seems like a fitting reward for a man who had already

distinguished himself during the Second World War, installing and inspecting steel roadways on Mulberry harbours – the temporary, floating pontoons that were so crucial to offloading cargo and vehicles after D-Day.

However, Coughtrie did not invent the mole grip from scratch. It was inspired by the very similar vise-grip locking pliers, which were created by Danish immigrant William Petersen in Omaha, Nebraska, in the 1920s. It seems likely that Coughtrie encountered vise grips during the war, and developed a new release lever in the 1950s to avoid infringing the patent on the latest version of the US-made tool.

Mole grips are especially useful when drilling, to lock whatever you're drilling in place.

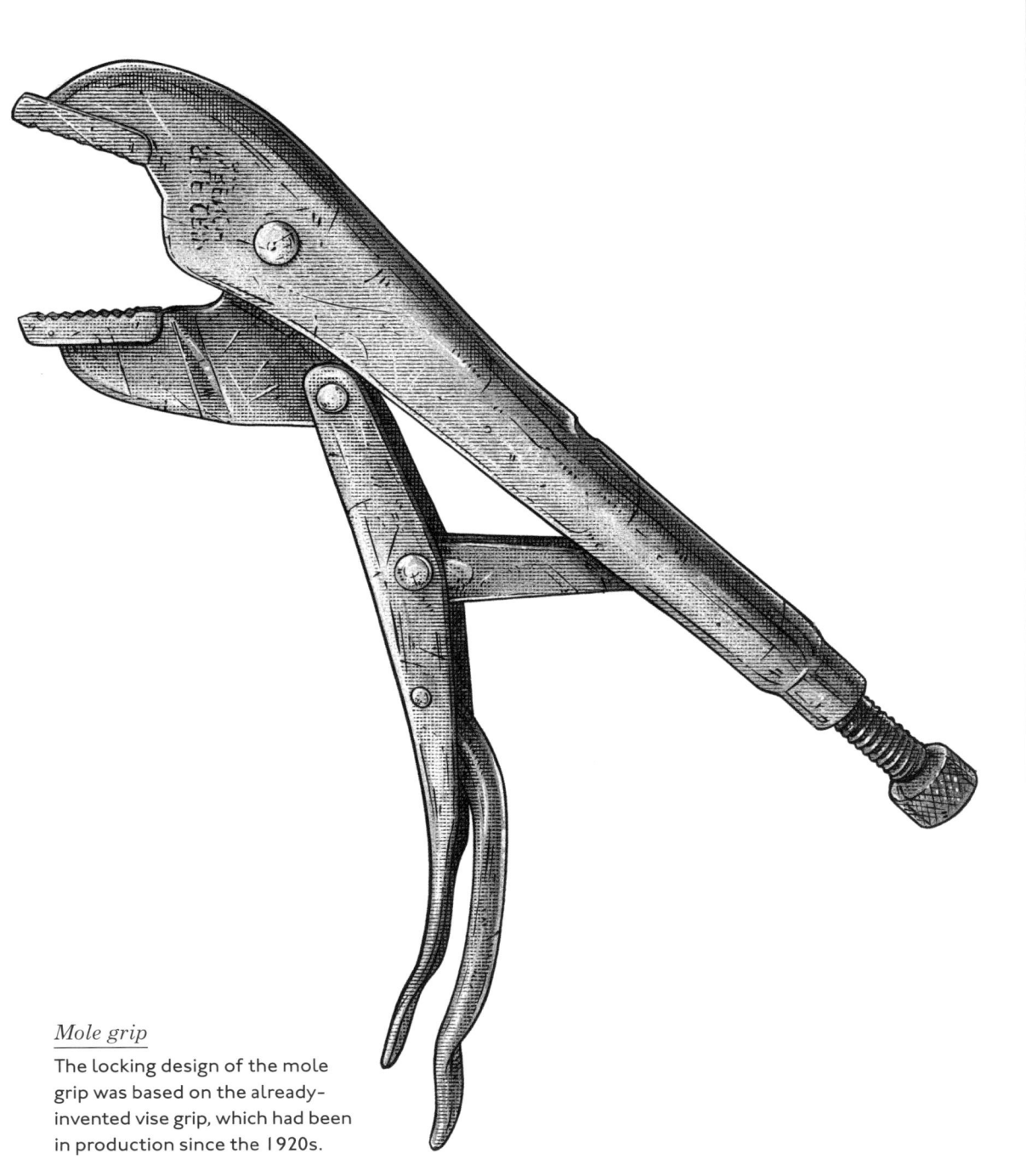

Mole grip

The locking design of the mole grip was based on the already-invented vise grip, which had been in production since the 1920s.

SLIP-JOINT PLIERS

Slip-joint pliers are so called because they have a pivot point that you can slip or slide to one of several positions depending on how wide you want the jaws to open. The most familiar type of slip-joint pliers are also known as Chanellocks or tongue-and-groove pliers, and you'll most likely know these ones because the jaws are offset from the handles at an angle of around 60°.

These pliers are versatile tools and are useful for gripping a whole range of items, as well as loosening and tightening old and misshapen nuts and bolts. I've always got a long-handled pair of Chanellocks to hand because they're reliable and get the job done without any fuss.

Slip-joints are thought to have been invented in the early 1930s by Howard Manning, the chief engineer at US hand-tool company Champion De Arment. The company was born from humble beginnings in 1886, with Pennsylvanian blacksmith George B. De Arment making farrier's tools and moving from place to place selling them from his wagon. The story goes that he'd spend winters making tools so that by spring, once the ice on the roads had melted, he'd be able to sell them. They had a particularly up and down 10-year period starting in 1893, when all their cash assets were stolen by a dodgy bank teller. But they bounced back quickly, and by 1902 the company, then known as the Champion Bolt and Clipper Company, was making a set of farrier's tools for King Edward VII.

Back to Howard Manning, who'd named his invention "Channellock" and was granted a US patent and trademark protection for it in 1934. Manning clearly wasn't happy about anyone trying to copy what he was up to (fair enough!) and trademarked his own shade of sky-blue for his grips in 1956. His Channellock pliers became so popular that the company even ended up changing its name to Channellock Inc. in 1963. The company is still owned by the De Arment family, and their tools are made in Meadville, Pennsylvania, where they've been based since 1904.

Howard Manning, the inventor of the slip-joint pliers, named his invention the "Channellock".

Slip-joint pliers

As the name suggests, slip-joint pliers have a pivot point that can slip or slide into several positions.

Tool No. **32**

DUCK-BILL PLIERS

These pliers have thin, flat jaws that look like a duck's bill. They're similar to needle-nose pliers (opposite), but their gripping surface, which can be serrated, smooth or cross-hatched, is larger. The jaws themselves can vary in length, as can the handles, which are sometimes much longer so you can access hard-to-reach areas.

I've got a pair of smooth-surfaced duck-bills that I borrowed from Pete Woods, the musical instrument and percussion restorer at *The Repair Shop*. He uses them a lot because duck-bill pliers are a really helpful tool for gripping and twisting wires and thin metal sheets. That makes them massively useful for jewellery makers and watchmakers, who often need to manipulate thin wires. They're also used by aviation and car mechanics for twisting safety wire and for gripping and bending cotter pins, which are used to secure rotating parts such as propellers and rotor heads.

One of the things I find them really useful for, which probably isn't exactly what they're designed for but is really handy, is if you've dropped something small or delicate, such as a washer, into a little hole. When you try getting the object with needle-nose pliers, it'll usually grab but then slip away from you at the last second. Just like those grabby claw machines you get at seaside amusements, when you think you've picked up a prize but then it slips agonizingly out of your grasp. But duck-bill pliers have that flat end that is much better at picking up awkward things.

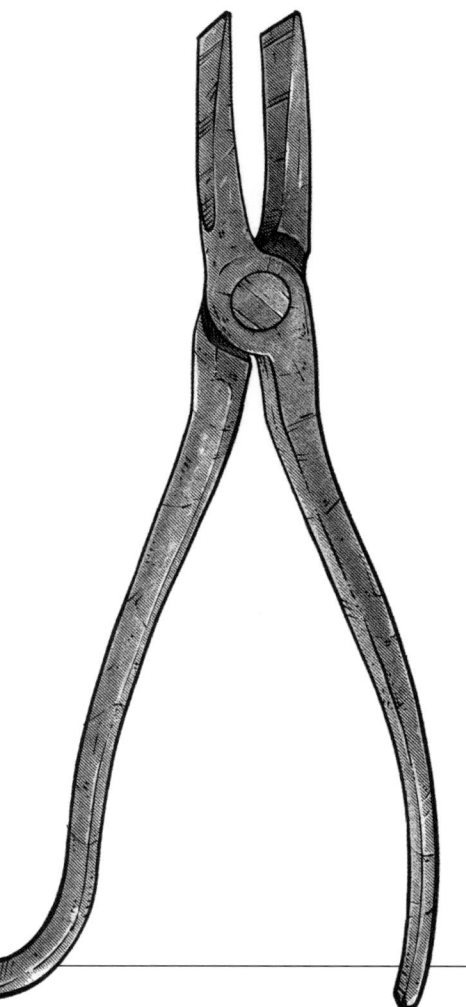

Maybe I'll start working on a version of that classic amusement game but with a duck-bill claw, so you might actually win something.

F.E. Lindström is one of the best and oldest manufacturers of duck-bill pliers, as well as other types of pliers and wire cutters. The company was founded in 1856 in Eskilstuna, Sweden, about 113km (70 miles) west of Stockholm, where it's still based. For a time it also made beautiful chisels with birch handles and forceps, tongs and even coffee grinders, but stopped producing all of these tools by about 1940 to concentrate on pliers and wire cutters. As of 1999, it's now part of Snap-On Incorporated (which also makes really good pliers).

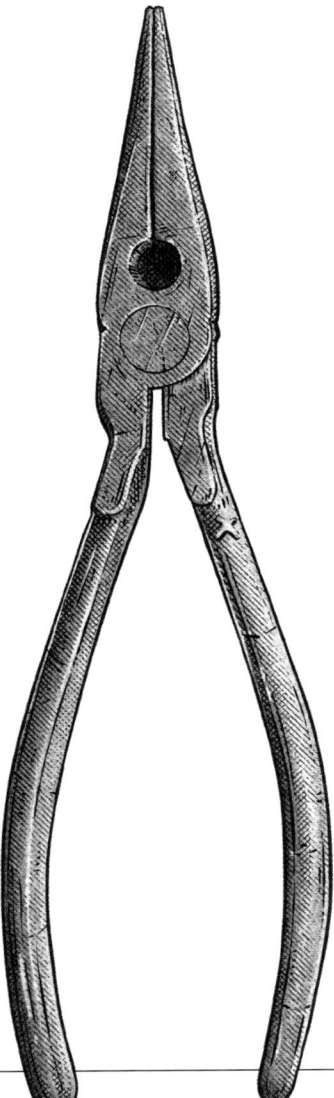

Tool No. **33**

NEEDLE-NOSE PLIERS

Also known as long-nose pliers or snipe-nose pliers, these pliers have long jaws that end in a fine point. They're useful for gripping small objects and holding, twisting and bending wires, which means they're essential for electricians, jewellers and clock repairers like Steve Fletcher at *The Repair Shop*. They've also got cutting blades for snipping through wires and cables.

One thing they're really handy for is picking up little things stuck in awkward places, such as behind the radiator. They're like a useful extension of your fingers when you can't get a proper grip on something. Whenever someone opens the door to my barn at *The Repair Shop* and asks if I've got a pair of needle-nose pliers, my first thought is, "What have you dropped down behind the radiator?!"

You can get quite a variety of needle-nose pliers – some feature ultra-lengthened handles or have bent heads, both of which are useful for mechanics working in awkward spaces. Some of the specialized needle-nose pliers for electricians even have insulated handles so they can work on 1,500-volt live wires – someone's got to!

CHAPTER 3
HAMMERING

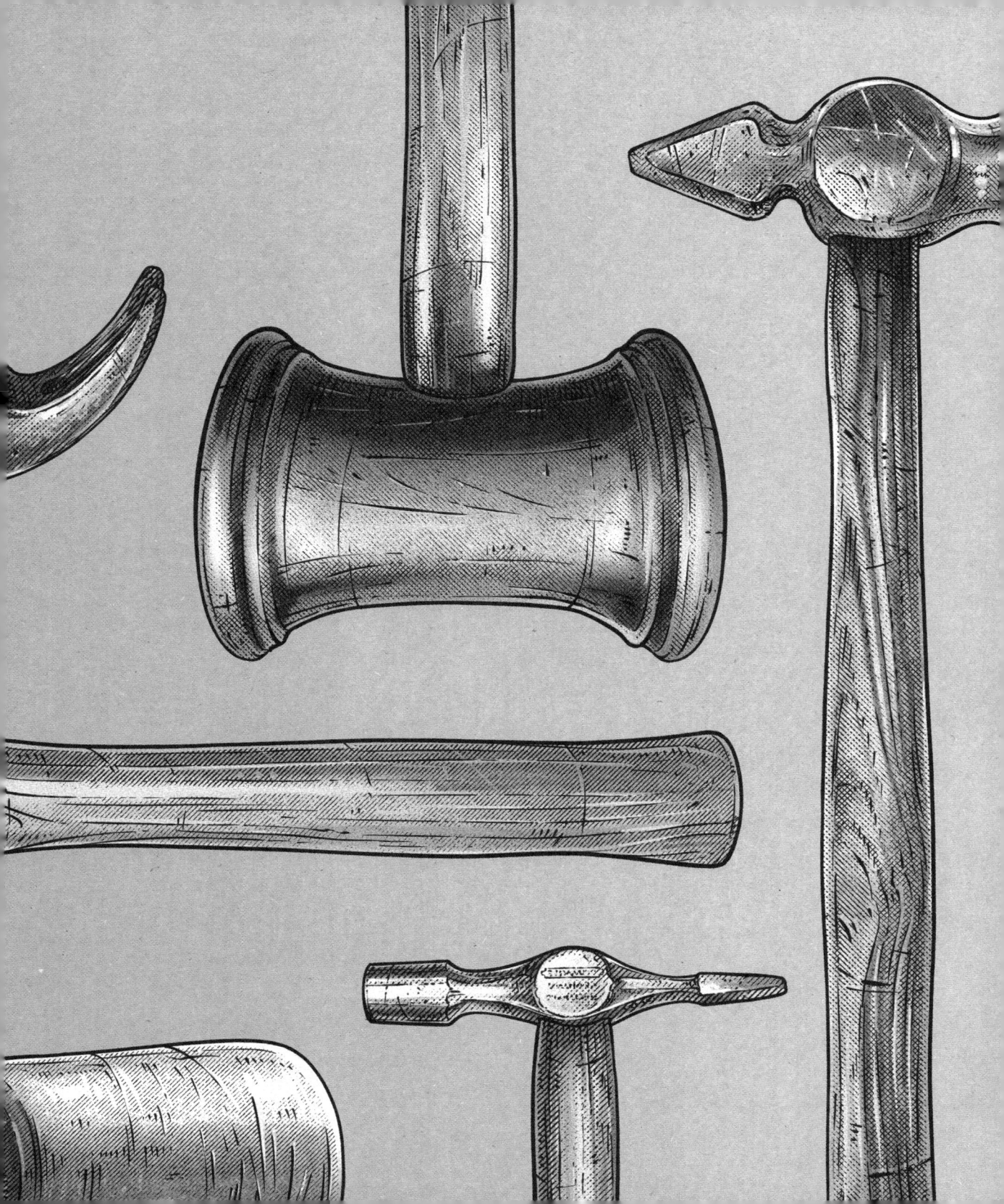

HAMMERS

In 2011 archaeologists discovered a site near the shore of Lake Turkana in Kenya, where they found stone tools, including hammerstones and anvils, that date back a frankly mind-boggling 3.3 million years. The hammerstones were made from sandstone or quartzite and were oval-shaped so they could be gripped firmly. They would have been used to strike other stones, such as flint, into sharp-edged flakes, which then became cutting tools.

When the first modern humans emerged between 45,000 and 30,000 years ago, they created new types of tools and, perhaps more importantly for the sake of the humble hammer, they developed handles. These would have been simple but sturdy wooden sticks or bones with leather or sinew tightly wrapped around the hammerstone and the handle to hold it in position. It would be another 20,000 years or so until bronze replaced stone as the material of choice for a hammer. Next it was iron, then steel. The idea of piercing the head of the hammer with a hole for a handle seems to have started in Europe during the Iron Age.

Since then, dozens of different hammers have been invented to perform every striking action you can imagine. But most now feature a head made of steel, which has been hardened using heat treatment, and a strong enough handle to stand up to heavy, repetitive force. When I think of a hammer, I can't help but think of a wooden handle, and that wood is often hickory, which is especially plentiful in North America but also native to Asia. The word "hickory" is a shortened from of a Native American word thought to refer to a drink made by mixing the ground nuts of the tree in boiling water. And hickory is a great choice for a handle, given that it's tough, durable, absorbs shocks really well and has a straight grain, which makes it easy to work with.

Tool No. **34**

CLAW HAMMER

This is the hammer most people think of when someone says "pass me the hammer". And they've been around for a fair while, too, dating back to at least ancient Rome.

The claw hammers used by the Romans were very similar to the ones you pass to your mate when they're putting a picture up. That's because the claw part of the hammer, with its V-shaped slot, is a simple yet genius design for levering out bent nails – a problem that even the Romans, with their amazing ability to build pretty much anything, would have come across every day. It's a comforting thought, imagining a toga-clad Roman saying the equivalent of "Oops, I've ballsed that nail up. Let's try again," before yanking the nail out with his claw hammer.

Fast-forward to Oldbury, a few miles west of Birmingham, and the year 1805, when some beautiful tools were being made by William Hunt & Sons at Brades Steel Works. They went on to produce an amazing variety of tools under their trademarked brand names Brades and WHS (the initials of the company name), including claw hammers. Many of these feature strapped handles, which was one attempt to solve the hammerhead-flying-off-the-handle problem by securing the head to the handle with rivets. On the subject of fastening the handle to a claw hammer, it was sometime in the late 1830s that the "adze-eye" was developed – a deep socket into which the wooden handle fits.

One important variation of the claw hammer is the framing hammer. Firstly, the head of a framing hammer is oversized and heavier, and the handle is longer, which is all designed so carpenters swing it less than they would do a regular claw hammer. Also, the claw is much straighter (although still slightly curved) because it's used more like a crowbar than for pulling out nails. The face is also checkered (milled) to reduce the number of glancing blows when you don't hammer the nail directly straight. And we're all guilty of that from time to time.

BALL-PEIN HAMMER

The ball-pein hammer, also known as a machinist's or engineer's hammer, has one flat face and one rounded (called the peen) and is used a lot by metalworkers. In fact, it's my go-to hammer, and I own more ball-peins than any other tool (and that's saying something).

Ball-pein hammers come in a variety of different sizes and shapes – weight-wise, they vary from under 100g (3½oz) up to about 1.5kg (3lb 5oz) – and I've used all of them at some point! Many of the ones in my workshop I've adapted to a particular job, such as shaping an internal curve of a pot. And if something similar comes up in the future I know which hammer to reach for. I remember using my biggest ball-pein hammer as a dolly (see page 158) on a *Repair Shop* job because the end was just the right rounded shape.

In case you're wondering what the "pein" part is about, peening is basically the process of hammering a surface to make it stronger, and that's one of the main things you use this type of hammer for. The flat face can be used for all sorts of tasks but is

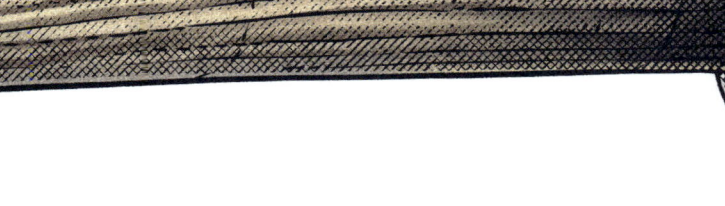
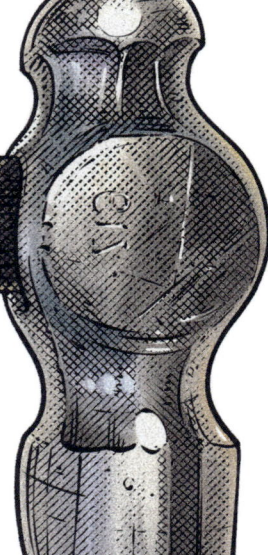

Ball-pein hammer
Look out for vintage ball–peins stamped with the names Gilpin or Whitehouse, who were both makers of these hammers.

most often used for striking chisels and punches. You also use the peen to set rivets in place and for rounding off the edges of pins and fasteners.

It isn't known exactly when ball-pein hammers came about, although there are many stories. What we do know is that one of the better-known 19th century producers of these hammers was William Gilpin & Son, based in Cannock, near Stafford. The company can trace its history back to 1763, but sometime in the 19th century, it became celebrated makers of augers, a drill-like hand tool with a crank-shaped handle that have been around in one form or another for centuries. One of the company's employees, a chap by the name of Cornelius Whitehouse, left in 1881 and set up his own business as an auger and edged-tool maker, which continued until 1967. I've mentioned these two because if you're looking for vintage tools, two of the more handsome ball-pein hammers you'll come across will be stamped with the names Gilpin or Whitehouse.

Tool No. **36**

PIN HAMMER

This lightweight hammer weighs only about 100g (3½oz). It's like a miniature version of a cross-pein hammer (see opposite) and is mainly used for driving pins, tacks and small nails into woodwork.

To use a pin hammer, start by gently tapping the pin or tack with the small, wedge-like end and then use the other end to drive it all the way in. If you're careful you won't whack your fingers!

These hammers are also used a lot by craftspeople who work with small pins, such as upholsterers attaching fabric to furniture frames, as well as carpet layers and picture framers.

STONEMASON'S HAMMER

One of the oldest trades in existence, stonemasonry is responsible for some of the most incredible buildings and monuments ever created. The Great Pyramid at Giza is the only one of the Seven Wonders of the World still standing, and it's the oldest of the seven by about 2,000 years.

To shape the roughly 2 million stone blocks needed to create the Great Pyramid, ancient Egyptian stonemasons would have most likely used mallets and chisels. This technique is still used to cut stone today, although the job has been made easier by the invention of the stonemason's hammer (also known as a brick hammer), which combines the functions of a mallet and a chisel into a single tool.

A stonemason's hammer has two faces: a chisel blade and a more traditional hammer face that's square instead of rounded. With the square hammer face, you can break up brick and stone, while the chisel blade is used for more delicate work, such as accurately chipping off pieces of stone.

One of the oldest stonemason's hammers I've seen was made by C. Drew & Company, a famous toolmaker founded in the coastal town of Kingston, Massachusetts, in 1837 by Christopher Prince Drew. He came from a long line of shipbuilders, six of whom made the tools, ironwork, bolts and braces needed to build the USS *Independence*, the first "ship of the line" (a big warship designed to fight in a line alongside other such ships) built by the newly formed US navy. Christopher's

business was famous for its caulking irons, which were used to drive cotton or rope fibres into the seams between planks of wood on ships to make them watertight. C. Drew & Company also made several high-quality masonry tools, and the company was in business until 1970. I don't own a stonemason's hammer, but if I ever see a C. Drew one at a boot sale, it's going in the van!

PLANISHING HAMMER

Planishing involves striking metal lightly against a curved surface to shape and smooth it. One face of a planishing hammer is usually flat, which smooths the metal, while the other is a rounded face, which shapes it.

Planishing hammers are used as a finishing tool on items that have already been worked on to create the general shape you're after. I use a planishing hammer all the time when I'm hammering sheet metal. They're often used by jewellery workers and panel beaters, who repair damaged and dented body parts of cars.

A planishing hammer is often struck against a handheld dolly (a small anvil – see page 158) or a planishing stake, which looks a bit like a mushroom head and matches the curvature of the metal item you're working on. The head is connected to a rod that is held in place in a vice or in the ground.

You can even get pneumatic (air-powered) planishing hammers that strike in excess of 1,000 blows per minute, which massively speeds up the process. But they're no good when you're dealing with very small pieces or you're working in a confined space.

"Planishing" means to shape, finish or smooth a metal surface by lightly hammering.

SLEDGEHAMMER & CLUB HAMMER

A sledgehammer is a no-nonsense tool with a dense, heavy, double-faced head and a long handle designed to give it momentum when you swing it. You use it with two hands, and you'll need two hands because heavy-duty sledgehammers can weigh well over 7kg (15lb 7oz). So what's it got to do with sledges? Nothing. It's thought that the word "sledge" comes from an Old Norse or early German word, which translates roughly as "striking forcefully". And that's exactly what you use a sledgehammer for: demolishing things, such as walls, or breaking up tough surfaces such as concrete.

Sledgehammers were (and are) traditionally used by blacksmiths, with one person holding the tool that's being worked on in position with a pair of tongs, while the other swings the sledgehammer and strikes a number of blows to shape the metal. The amount of trust these two people need to have for each other is significant, because one little move this way or that and they're going to be in trouble!

A club hammer (also called a lump hammer) is the smaller brother of the sledgehammer. It weighs around 1–2kg (2lb 4oz–4lb 8oz) and is designed so you can swing it with one hand. They're often used for smaller-scale demolition work but also work well when striking chisels and driving nails into masonry. They're also popular with some carpenters and furniture makers for heavier woodworking jobs.

A club hammer is a smaller, lighter version of a sledgehammer.

Club hammer

Club hammers usually have two identical faces and a shortish handle. The matching faces balance the weight of the hammer, making it easier to use.

Sledgehammer

Sledgehammers have longer handles than club hammers and are better suited to heavy-duty demolition work.

CROSS-PEIN HAMMER

A cross-pein hammer has two faces, one slightly convex and the other wedge-shaped. It's called a "cross-pein" because this wedge-shaped end is crosswise to the direction of the handle. It's basically a general-purpose woodwork hammer used mainly for driving in nails and tapping joints together, but it's also used by metalworkers for stretching metal.

You see the company names I. Sorby and Brades come up quite a bit with vintage cross-pein hammers. We've already mentioned Brades (see page 63) but I. Sorby is worth going into. I. Sorby was the mark used by Sheffield tool producer John Sorby from around 1810, who formed a partnership with John Turner and Henry Skidmore a few years later. Sorby and Skidmore died in the 1820s, but John Turner continued making tools and the rights to I. Sorby passed to his sons Charles and Joseph Turner in 1854. The Turners continued to use the I. Sorby mark due to its reputation; it was very well respected.

Cross-pein hammer
You might be lucky enough to find a vintage cross-pein stamped with the "Punch" trademark, of Punch and Judy's Mr Punch.

In 1859 the Turners registered the "Punch" trademark, which featured Mr Punch (from Punch and Judy fame), and began to stamp it on their tools. It was used all the way up until 1963, when the company was dissolved and the factory demolished. In 1871 the company bought a new site – Northern Tool Works – and William Marples, owner of his own successful tool business, which he'd started in 1830, bought an interest in the company for his son Charles. Soon, the company was employing over 80 people, and it became one of the largest tool factories in Sheffield. But the company lost its way after Charles left, and in 1909, after going into financial trouble, the firm was taken over by William Marples & Sons.

William Marples went from strength to strength, and by the outbreak of the First World War, they were employing 400 people and supplying tools for carpenters, engineers, masons and plumbers. But, after having being owned by the Marples family for generations, in 1962, the business was bought by C. and J. Hampton (maker of Record tools) and W. Ridgeway & Sons, who we'll go into later.

You'll come across the name W.M. Marples & Sons and from 1875, their trademark Hibernia (after their factory, Hibernia Works) and their triple shamrock mark if you're searching through vintage tool listings on the Internet. And I do that a fair bit.

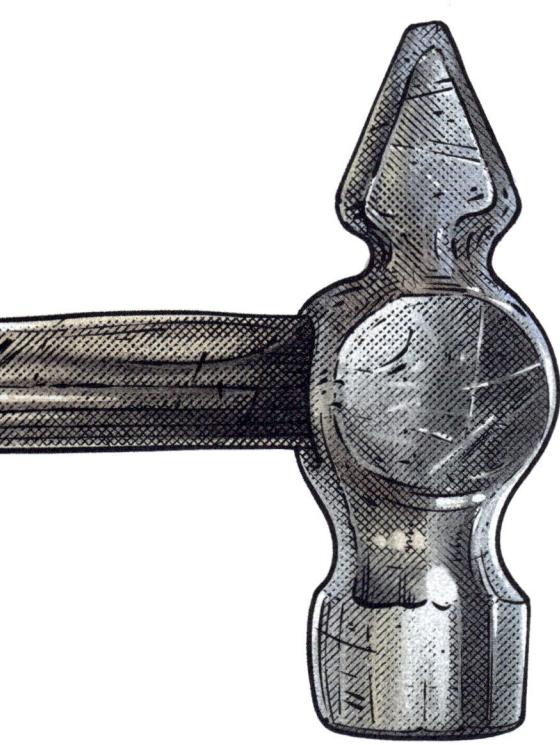

Tool No. **42**

BLACKSMITH'S HAMMER

Size-wise, a blacksmith's hammer is similar to a club hammer (see page 68), but one of its ends forms a wedge shape that's crosswise to the direction of the handle. For this reason, it's technically a cross-pein hammer but, as you can see, it doesn't look much like the sort of cross-pein hammer you'd use for woodworking (see page 70).

A blacksmith's hammer is much heavier and denser than a club hammer, and it needs to stand up to repeated hammering on metal. One of a blacksmith's hammer's main uses is for forging, which involves heating metal in a forge, holding it with a pair of tongs over an anvil and then hammering it into the shape you want. Blacksmith's hammers are also used for riveting, which involves joining two pieces of metal together with a rivet; a short, straight metal bolt with a head at one end. The rivet is hammered into place and the tail end is also hammered repeatedly to create a second mushroom-like head, so the rivet stays in place.

If you're wondering where you've seen the distinctive shape of the blacksmith's hammer, it might be because it appears in the gold hammer-and-sickle emblem on the flag of the former Soviet Union, against a red background. The hammer represented the industrial workers, and the sickle symbolized the rural peasantry, so together they stood for the unity of the working class.

I found an old blacksmith's hammer head in a plastic bag full of old rusty tools that someone had dropped off at the set of

The Repair Shop. These things have a habit of ending up at my door, and for good reason, because I usually find a home for them. Although, I wasn't sure what to do with this hammer head – it was too big and heavy for panel beating, and it wasn't ideal for general hammering. But I wondered if Nick, who volunteers at the forge in the Weald & Downland Living Museum, wanted it. When he saw it, it was like he'd just had a visit from Father Christmas!

Nick put a nice handle on it straight away and it's now become his favourite hammer. He uses it every day, and it makes me smile every time I see him with it. It's great to pass a tool on to someone who you know will love and take care of it. I'm a big believer in karma, and some of my favourite tools are ones that have made it to me in a similar way. It reminded me that sometimes you really can find a diamond in the rough. Even the most horrible-looking rusty, pitted hammer can come back to life if it's shown a bit of love and attention, and can become useful once again.

The distinctive shape of the blacksmith's hammer appears in the hammer-and-sickle emblem on the flag of the former Soviet Union.

The first person to develop a mechanical hammer for the purpose of beating gold was none other than Leonardo da Vinci.

Tool No. **43**

GOLDBEATING HAMMER

Goldbeating isn't just a case of using the same hammer and bringing it down thousands of times in exactly the same way. It's more complicated than that, and involves using hammers of different weights struck with varying amounts of force.

The first thing you notice about any goldbeating hammer (before you even pick one up) is that it looks really heavy. They're made of cast iron and either look like a bell or a dumbbell. The heaviest of them is used at the beginning of the job, and this hammer can weigh as much as 8kg (17lbs), which, just for the record, is more than a heavy-duty sledgehammer (see page 68). The other, lighter hammers typically weigh between 3 and 6kg (6 and 13lbs). And let's remember that while a goldbeater is bringing down this massive bit of metal, their fingers are centimetres away from the gold leaf, which they're constantly turning.

Goldbeating has remained relatively unchanged since medieval illuminated manuscripts were decorated with gold as far back as the 11th century. Due to the time, force and strength needed to beat gold, gold leaf is very rarely completely handmade. So, the first beating stage is almost always performed by a tilt (or trip) hammer, a mechanical hammer with a pivoted lever. The first person to develop a mechanical hammer for the purpose of beating gold was none other than Leonardo da Vinci, but like his flying machine, there's no evidence it was ever built. In the end, it wasn't until 1928 that a machine was created that reproduced the traditional motions of goldbeaters. Fast-forward, and in factories producing gold leaf today, we're talking about hammerheads weighing over 600kg (1,300lb). There are very few traditional gold-leaf workshops still around, but you'll find that a small power hammer is the only machinery used. Everything else is done, painstakingly, by hand, and the quality you achieve this way is hard to beat.

RAWHIDE MALLET

A rawhide mallet is made from tightly rolled up cow or buffalo leather that is shaped into a cylinder and screwed on to a wooden handle. When you look at the end, it has that Swiss roll-kind of effect.

Heavier rawhide mallets have a core of steel or lead. The outer surface of the hide is usually coated in a varnish, lacquer or shellac, so the surface is hard to begin with, but you can condition it like you would with a pair of new leather shoes. Soaking the mallet head in warm water and then hammering the ends against a hard, smooth surface softens up the leather and means you won't get any debris stuck in the hide. The leather will continue to soften over time, but it's a durable material.

Rawhide mallets are essential tools for jewellery makers, leatherworkers, instrument makers and other metalworkers because they allow you to shape and form metal without leaving scratches or causing dents. It means you can deliver a softer, quieter and more controlled impact than with a regular metal hammer. In jewellery making, it's used together with a triblet (a tapered, thin cylinder) for shaping and stretching rings. In leatherworking, a rawhide mallet is great for striking other tools, such as bevelers and stamps, used to make impressions in the surface of the leather.

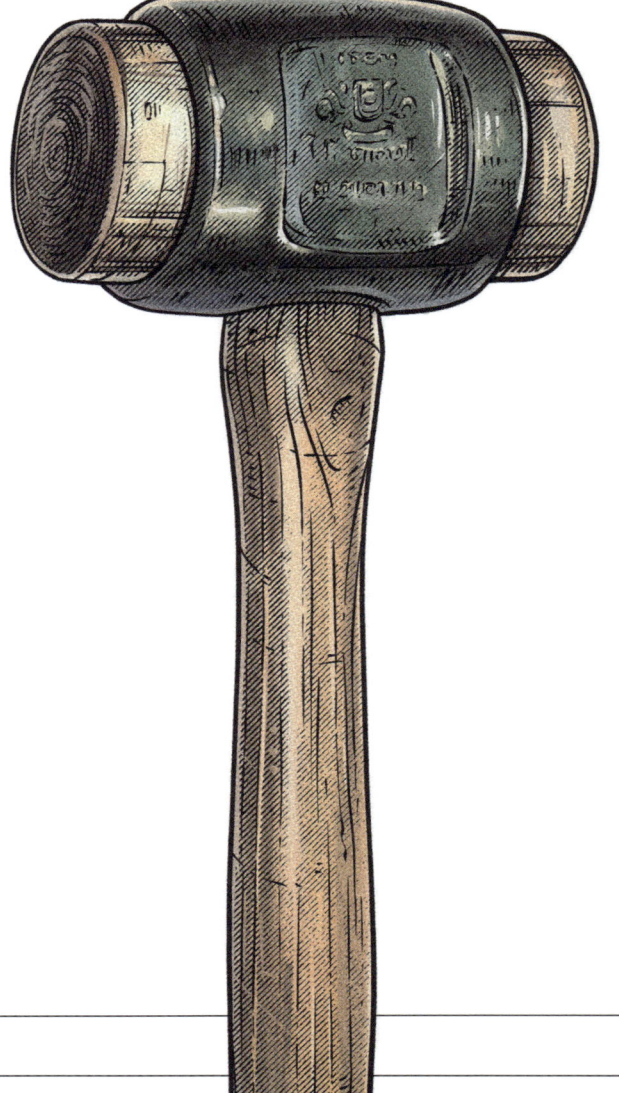

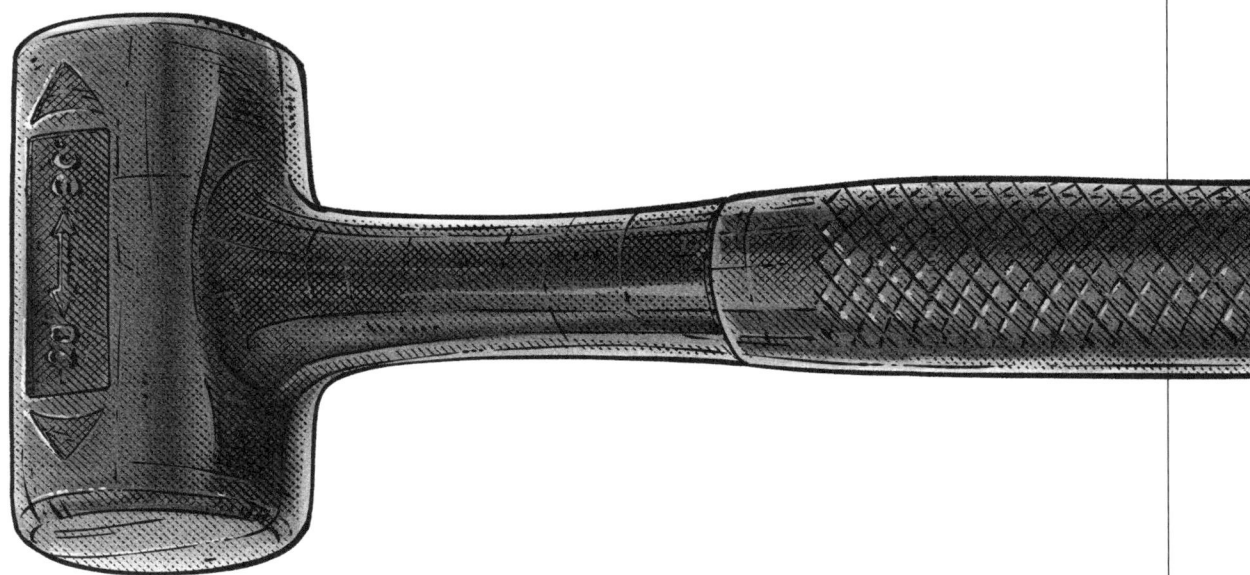

Tool No. **45**

DEAD BLOW MALLET

The name sounds like some sort of fearsome war hammer intent on causing as much damage as possible, but this type of mallet is actually designed to prevent it rebounding against or damaging the object you're striking.

Dead blow mallets have a cylindrical head usually containing thousands of steel pellets or lead shot. So, unlike a regular hammer, when the head of a dead blow mallet strikes an object, not all of the mass hits it at the same time – the impact is spread out and the force feels deadened, hence the name.

Dead blow mallets are usually made from one continuous piece of moulded plastic or rubber. They are similar to rubber mallets, but a rubber mallet is lighter, usually has a solid rubber head without anything inside it and has a wooden handle. That means they tend to rebound if you're not careful!

They're often used by mechanics when they need to deliver concentrated force without damaging other, nearby working parts, and for hammering out small dents in the bodywork of cars. They're also used by carpenters for tapping together tight woodwork joints.

CARVER'S MALLET

Unlike most mallets, which have a square head, a carver's mallet has a rounded head and is sometimes made from a single piece of wood (you can even make one yourself if you've got a wood lathe).

For these reasons, carver's mallets are quite beautiful and look more like a piece of finished furniture than a tool. The rounded face is designed to increase the amount of surface area available to strike a chisel or a gouge compared with a square mallet, and means that you don't need to concentrate as much on exactly where the face is. They're also lighter than most other mallets, so they're ideal for precise, delicate work such as carving intricate details into wood. The fact that they're made of wood also means you'll avoid damaging the handles of the tool you're using to strike them with, and won't end up with your mallet rebounding off them.

Carver's mallets can be made of beech, ash or hornbeam, but some are made from more exotic woods, such as African blackwood or olive wood. They also come in all sorts of sizes depending on the amount of material you need to remove, but the best ones feel like an extension of your hand. The only problem is that they have a habit of rolling off the workbench and onto your foot.

You can also get brass carver's mallets, otherwise known as journeyman's mallets, which are small but strong and designed for finesse. You can either hold them by the handle or grasp the brass end with your fingers for really delicate work. They're also good for working in small spaces.

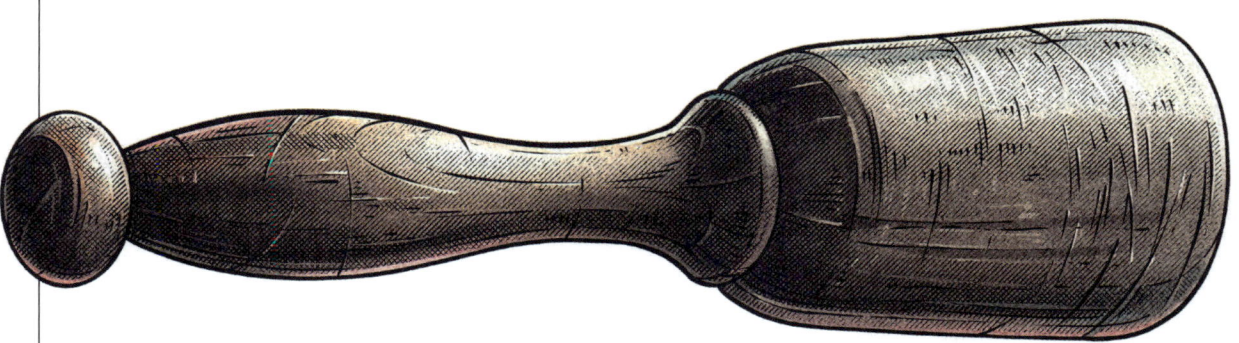

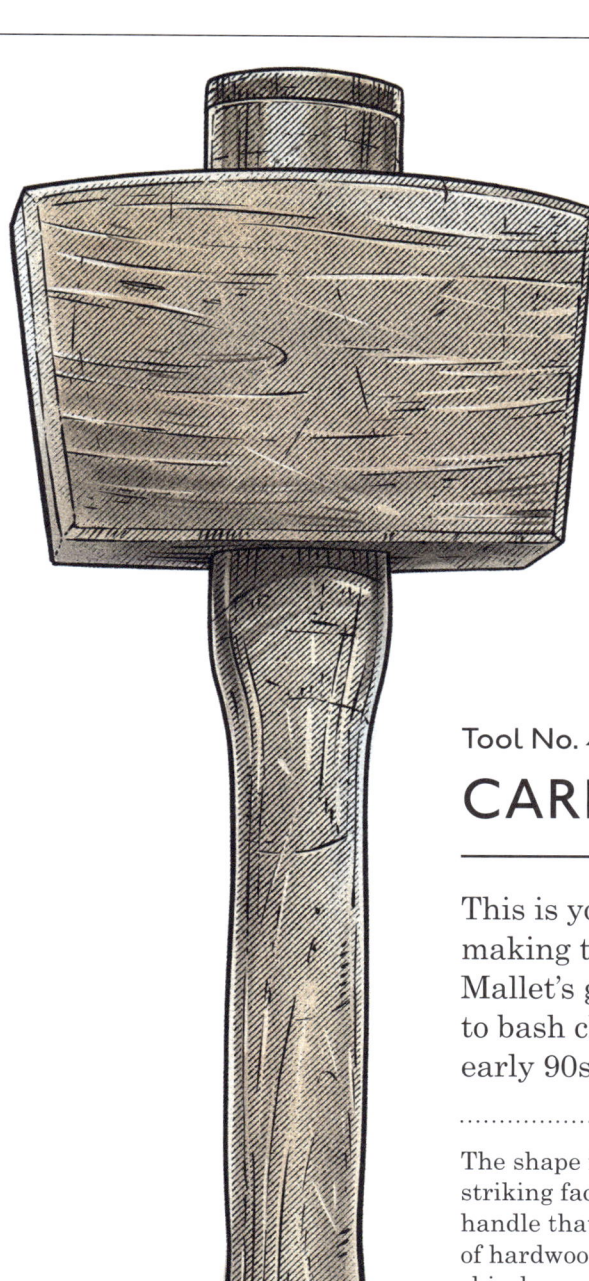

Carpenter's mallet

This is the classic shape we tend to think of when anyone mentions the word "mallet". These simple wooden mallets can be picked up quite cheaply, but it's also relatively easy to make one yourself.

Tool No. **47**

CARPENTER'S MALLET

This is your classic joinery and cabinet-making tool, and the inspiration for Timmy Mallet's giant soft pink mallet that he used to bash children over the head back in the early 90s.

The shape is instantly familiar, with its two tapered striking faces, large head size and the top of the handle that protrudes. Carpenter's mallets are made of hardwood, normally beech, and are used to tap chisels, gouges and other carving tools, as well as for banging panels and frames together.

CHAPTER 4
CUTTING

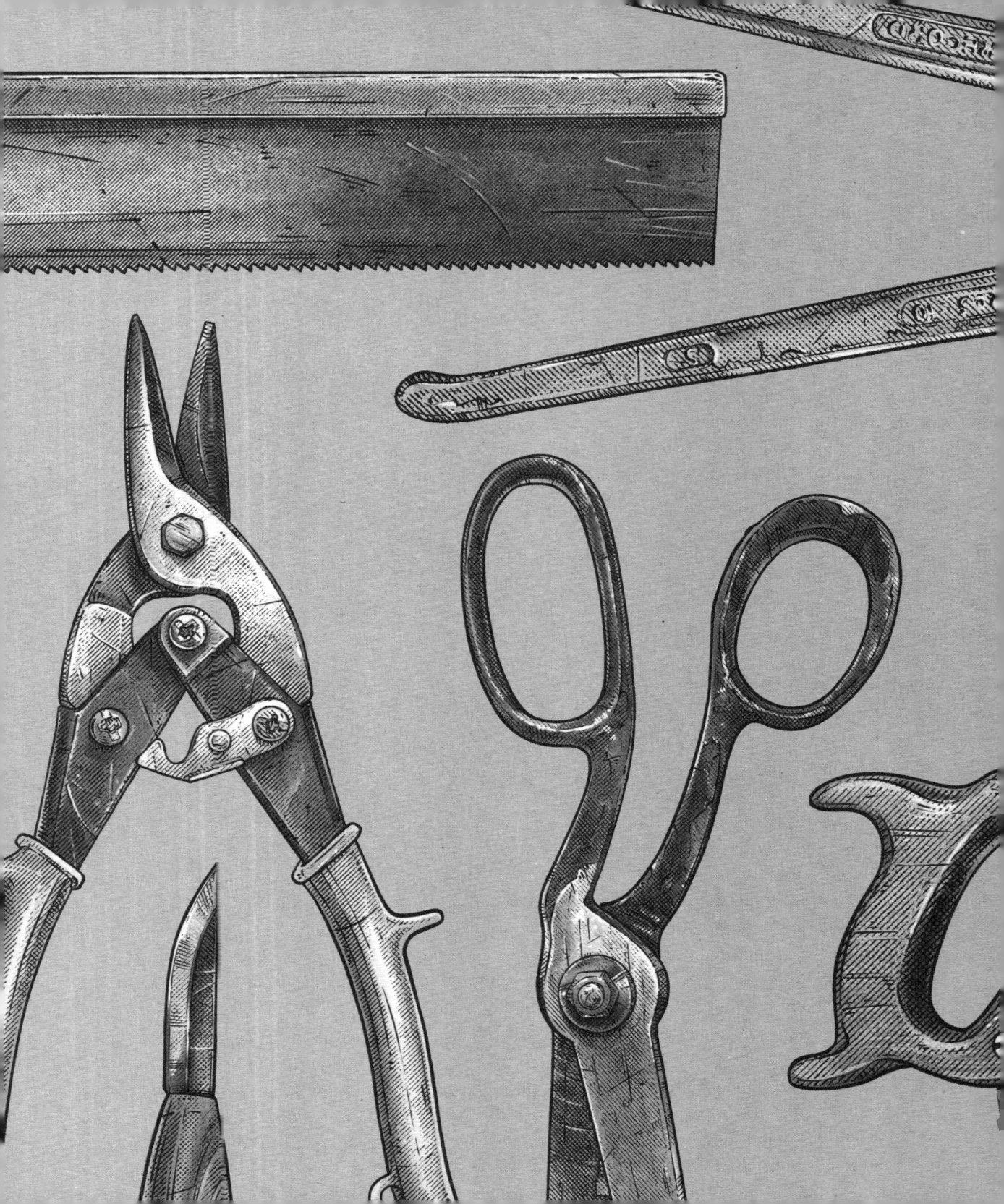

DRAW KNIFE

A draw knife is a traditional woodworking tool with one long blade and two handles positioned either side of it. It's used to remove slices or shavings of wood, and it's called a draw knife because of the way you use it: by pulling it (or drawing it) towards you with both hands.

You don't start at the end of the object when you're using a draw knife: you begin at the centre and move towards the end. It's an ancient-looking tool and has changed little since Viking shipwrights used them to shape and smooth their longboats and remove the bark from trees. And we know that because one of the oldest draw knives, dating from the 1st century CE, is in the Swedish History Museum in Stockholm.

Draw knives are often used by a woodworker sitting on a shaving horse, a sort of combination of a workbench/actual bench with a pedal that you press to clamp whatever you're working on into position. They're constructed entirely of wood and often made by the woodworker who's using them. It takes a lot of skill to use a draw knife properly, especially because the depth of the wood you're cutting is determined by the amount you tilt the blade. What has changed about it since the Viking days is the angle of the blade and the position of the handles, which would have been at a right angle to the blade. More modern ones can have curved blades and handles set at around 45°. Also, the bottom of the handles can sometimes be level with the blade or set below it. But whichever the angle and shape of the blade, after you've finished your work with the draw knife, it would have been on to the spokeshave, lathe or sanding block for finer work and smoothing.

This ancient-looking tool has changed little since Viking shipwrights used them to shape and smooth their longboats.

One thing that draw knives are still commonly used for is to make cricket bats. After you've cut your 75cm (30in) block of around 15-year old willow, it's left for about a year to remove any moisture in the wood. Then this piece of wood, known as the cleft, is cut into the rough shape for the bat's blade before going through a press to curve the blade's face and squeeze the fibres in the wood together. The next job is to cut out a V-shape out of the end of the bat, which is where the handle will attach. Then it's on to the most difficult work – shaping the profile of the bat – and that's the part that's still done with a draw knife. It takes some serious skill and experience to get this right. After that, it's on to the smoothing, sanding and polishing.

Draw knife

As well as being popular with the Vikings, draw knives were also used by medieval Russians for smoothing wood after it had been cut with an axe or adze (see page 117).

STANLEY KNIFE

The Stanley knife is a difficult one to put in a category, because, well, everyone (in the UK, Australia, New Zealand and parts of Europe, but not the USA – see below) just knows it as a Stanley knife. If we're being technical (which I suppose we should be), it's a utility knife with a retractable and replaceable blade.

The Stanley knife is used by all sorts of different crafts and trades, but often for cutting carpet, cord, packaging and cardboard. When they first made their utility knife back in 1936, Stanley referred to it as "the No. 199". It had a fixed blade but it was replaceable, and was made with a strong, die-cast zinc body. You can still get the No. 199, and I prefer it to the version with the retractable blade.

Frederick T. Stanley was born in Connecticut in 1802 and was one of seven kids growing up on a family farm. As a young man, he worked in all sorts of jobs, including as a clerk on a steamboat, a travelling peddler and a machine manufacturer. He went into business with his brother in 1831, making door locks among other things. The business didn't survive the big financial crisis of 1837, but they recovered and founded the Stanley Bolt Manufactory in 1842, and The Stanley Works in 1853. After a few good years, the business started diversifying, innovating and trying to sell its wares abroad. This was massively successful, and by 1919, sales were around $11 million (about $170 million today).

In the 1920s and 1930s, they began to acquire other companies, such as J.A. Chapman, a maker of hand tools based in Sheffield. During the Second World War, the company made over 30 million cartridge clips and 450 million machine gun-bullet belt links. It wasn't until 1961 that it updated its very successful utility knife, though, but this one featured a novel invention: a retractable blade, a design that is virtually unchanged today. Amusingly, Americans don't actually call this a Stanley knife. They call it a box cutter or utility knife.

Tool No. **50**

SLOYD KNIFE

A sloyd knife is used to carve wood. I always assumed Sloyd must have been a person, but it turns out the word comes from the Swedish *slöjd*, which means craft or handicraft. The word also refers to a unique approach to education that began in Finland in 1865 and soon spread to Sweden, where it was developed by a chap called Otto Salomon.

Salomon thought that school classrooms were boring places, that lessons were uninspiring and that sitting at a desk for hours on end without getting fresh air and doing some manual work was not helping kids to flourish. So in 1875, he started a school, but for teachers, not kids. The idea was that the teachers would learn handicrafts (especially woodworking) and go on to teach kids what they'd learnt. At its heart, Salomon believed that there was an important link between building things with your hands and intelligence, self-reliance, patience and even morality.

Salomon's idea was so popular and effective that it became part of the curriculum in Denmark, Sweden, Finland and Norway. It still is! The first tool the kids were given to work with was a carving knife, which became known as the sloyd knife.

Over time, the kids would build up their confidence and skill before moving on to make more complex objects. The things they'd make would be useful tools in their everyday lives, such as rakes, hammer handles, wooden cutlery, stools and chairs. All the while, their teachers were focusing on how the kids were developing.

As for the sloyd knife, Salomon wrote that it should be "made of good steel, about four inches long and not more than ¾ inch broad. The edge should be straight and two faces which form it should extend over the entire width of the blade... The blade ought not to taper to a point... The best angle for the edge is 15°." It's a great multipurpose knife, used for chopping, marking and whittling, so you can see why it was considered the foundation for sloyd education.

Tool No. **51**

MARKING KNIFE

A marking knife is a preparation tool that you use to cut a narrow, shallow line in wood that you then follow up with your cutting tool, chisel or plane. You use a marking knife together with a straight-edge or a square to guide your line.

Marking knives usually have a wooden handle and either a skewed (slanted) or spear-pointed blade, but some marking knives are made from a single piece of steel. A marking knife is also known as a striking knife, although old striking knives from the late 18th century don't look much like modern marking knives. They were made from a single piece of iron and have two very different ends: one with a thicker, flat section ending in a skewed blade, and the other tapering to a scribing point. So, they basically combine a marking knife with a scratch awl (see page 25).

Between the two ends are rounded sections and little ridges for your fingers to grasp the tool. One of the earlier illustrations of this type of knife appears in Joseph Smith's "Explanation or Key to the Various Manufactories of Sheffield with engravings of each article designed for the utility of merchants, wholesale ironmongers and travellers" from 1816. Lengthy title, I know, and for that reason it's better known as "Smith's Key" after the editor and engraver who put it together. It's one of the first illustrated catalogues of hand tools I'd ever seen and it's a beauty.

A marking knife is also known as a striking knife.

Tool No. **52**

EMBROIDERY SCISSORS

These scissors are difficult to confuse with any other type of scissors because they're shaped like a stork, with the blades forming its beak and the handle its legs. Some of them are stunning works of art, really, with fine details of a stork's feathers and feet incorporated into the design. So what's the stork about then?

If you guessed "something to do with childbirth", you're on the right lines. They were originally part of a midwife's toolkit, designed for clamping the baby's umbilical cord before it was cut with a pair of surgical scissors. Their blades were rounded and dull because they weren't intended for cutting. And they were shaped like storks because of the centuries-old connection between these birds and babies (no one knows the precise origin, but the best guess is that the idea took root in the Middle Ages because storks start migrating from northern Europe in the middle of summer, the season associated with fertility, and return nine months later ready to have their chicks).

As for the connection to embroidery, midwives would often spend a lot of time waiting for the birth to progress, so they'd usually travel to work with their embroidery boxes so they could fill the time. Since the 19th century, the umbilical clamp version of the scissors changed, becoming longer and with the blades turning into proper cutting tools, with sharp, fine points. But they kept their stork design in tribute to the connection to childbirth.

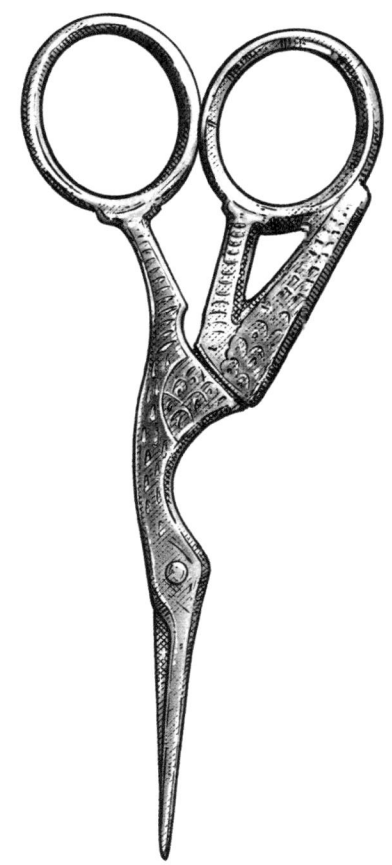

TAILOR'S SHEARS

For some crafts and professions, one tool rises above all others. For tailors, it's their "sidebent" tailor's shears, and the relationship between the two is a big deal. Some tailor's shears are passed down from generation to generation; others have been carefully chosen by an apprentice who will probably keep using them until they retire.

We know that tailors have been using shears to cut cloth for hundreds if not thousands of years, but the story of tailor's shears in the form that we know today goes back to the Steel City (aka Sheffield) in the middle of the 19th century. At that time, there were a handful of renowned scissor-makers working in the city, including Thomas Wilkinson. He came up with a genius design for tailor's shears featuring cranked handles that tilt upwards so that the lower blade runs flat along the fabric. And that means you can cut fabric much more smoothly.

We know that Queen Victoria and Prince Albert were fans of his work, because Wilkinson received a Royal Warrant in 1840. Thomas Wilkinson's son George followed in his father's footsteps and the business became Thomas Wilkinson & Son, and together they won a medal for their display of scissors at the Great Exhibition at Crystal Palace in 1851. The company was taken over by William Whiteley & Sons, another renowned Sheffield scissor-maker but with an even longer history than the Wilkinsons, dating back to 1760. They had their own royal appointment and medals awarded during the Great Exhibition, so they weren't messing around either. And they're still very much around today in Sheffield, with the 11th and 12th Whiteley generation running the company.

The blades of tailor's shears are really long so they can cut through a large area of material in one movement, which is great for accuracy and efficiency. Anyone who's ever cut Christmas wrapping paper with a small pair of scissors will appreciate them, I'm sure! Tailor's shears are heavy, but you shouldn't actually need to lift them much – they're designed so you can just slide them along a tailor's board to cut material.

Tailors actually have two pair of scissors or shears – one for cutting cloth and one for cutting paper (which they use to draw up their patterns). They absolutely don't use the cloth shears to cut paper – that would blunt them, and to be honest, that's probably seen as a criminal offence among tailors.

The blades of tailor's shears are really long so they can cut through a larger area of material in one movement.

PINKING SHEARS

Pinking shears are large scissors with saw-toothed blades that leave a zigzag pattern on cloth, although they're used to cut through other material too. They're not just used to create a pretty decoration, though – the design serves a useful purpose. Cutting a straight line through cloth can leave frayed edges, which can mean threads start to come undone. Cutting a zigzag pattern helps stop this from happening.

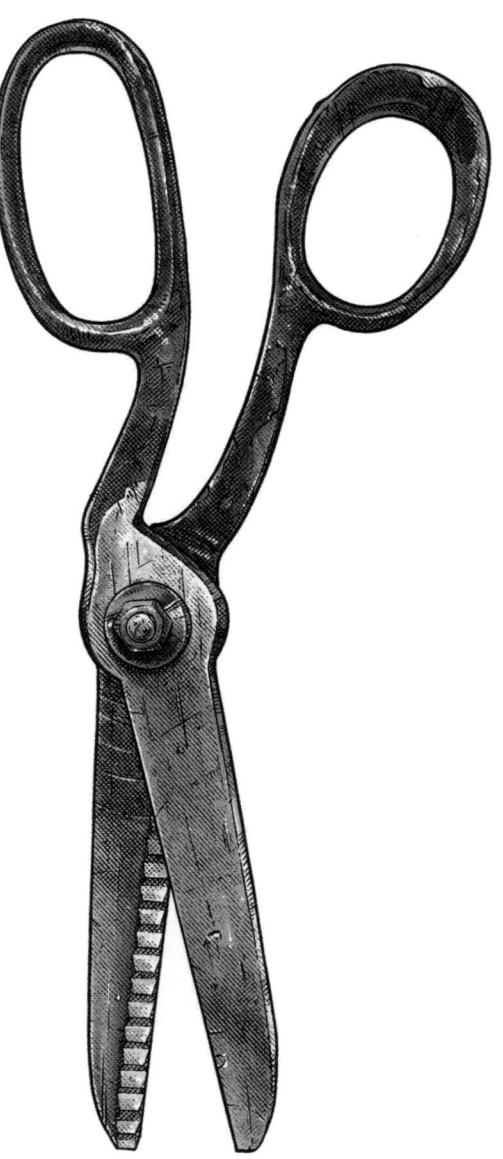

I know what you're thinking, what's the word "pinking" about? So, the verb "to pink" appeared in the 13th century, meaning to pierce with a pointed weapon. But sometime in the 16th century, another meaning emerged: to decorate an object's edge with a regular pattern of small holes. And it's this definition seems like the most likely origin for the word "pinking", meaning to cut a zigzag edge.

The first design for a pair of pinking shears was patented in 1893 by Louise Austin from Washington state, USA. Her patent application describes them as having "blades made much thicker than in ordinary scissors or shears, and provided with male and female matching, projections and depressions of any desired configuration on their inner broad faces, the projections and depressions running transversely to the length of the blades..." The application didn't just include the zigzag-shaped blades, but also six other designs for cutting edges.

WIRE STRIPPERS

This is a tool used mainly by electricians, to remove the protective insulation from electric cables without damaging the wire inside them. There are two main types of wire stripper: manual and compound automatic.

Manual wire strippers come in two forms, but both of them are adjustable to match the size of the wire. The first is pretty basic and typically has a screw and nut that let you adjust the width of the pincer-like jaws to match the size of the wire. The second type looks more like a pair of pliers or secateurs, and has several pre-formed holes, so you just work out how thick your wire is, feed it through the hole and squeeze the handles to cut through the insulation. Then you just slide the strippers to remove the jacket and expose the wire.

Compound automatic strippers are a bit cleverer (and pricier). When you squeeze the handles, sprung-loaded teeth work out the size of the wire, cut through insulation and remove the jacket, all in one motion.

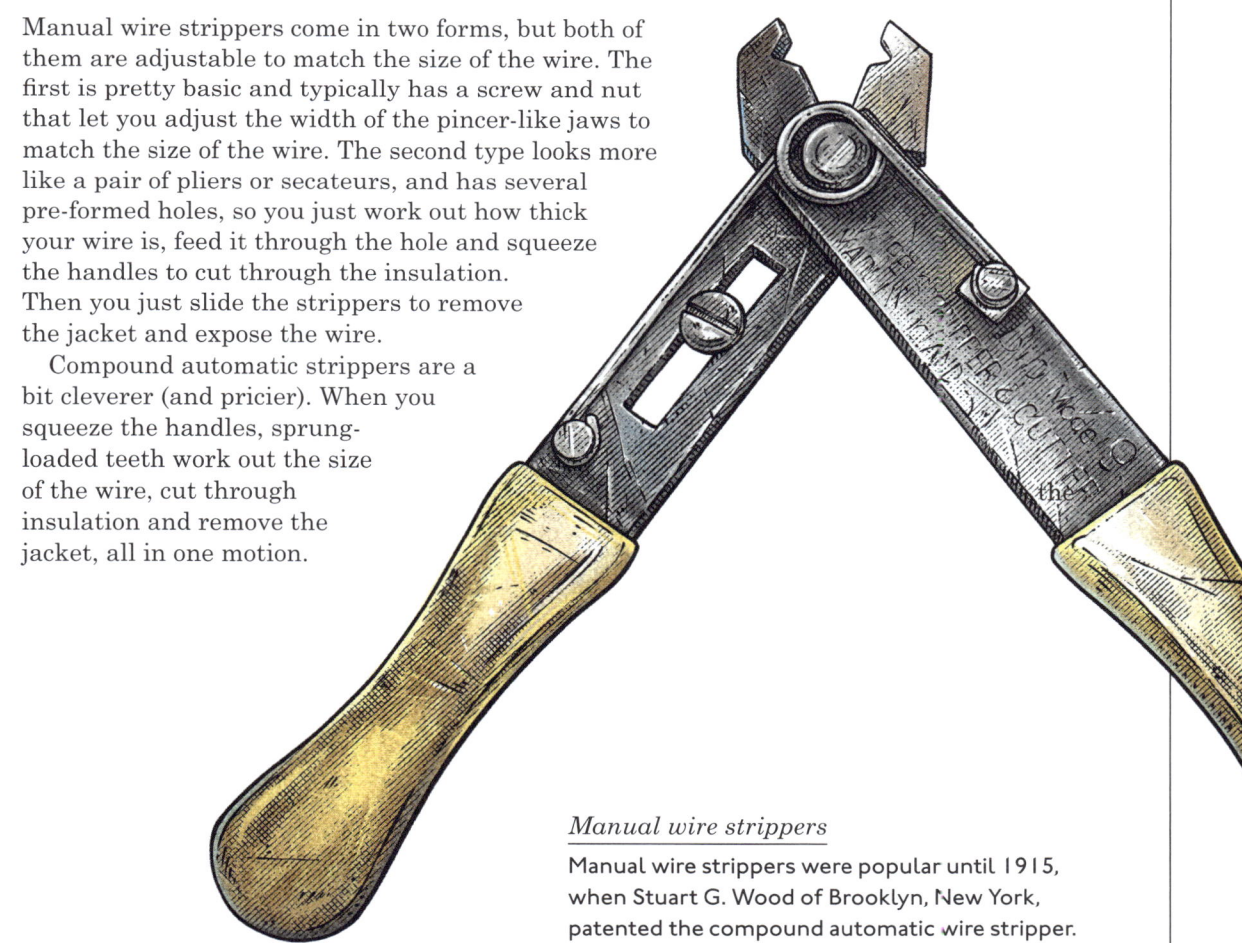

Manual wire strippers
Manual wire strippers were popular until 1915, when Stuart G. Wood of Brooklyn, New York, patented the compound automatic wire stripper.

Tool No. **56**

WIRE CUTTERS

Wire cutters go back to at least the 19th century, most probably originating in the USA around the time that barbed wire was invented in the 1860s. They've been associated with conflict ever since, including the Fence Cutting Wars in the American Old West in the 1880s, when barbed wire was used by established cattle ranchers to fence off land (including public land) to protect it from the claims of new settlers, who fought back with wire cutters to take apart the fences.

One pair of wire cutters with an amazing war story arrived on my workbench at *The Repair Shop* a few years ago. It begins with a 23-year-old British lance corporal by the name of Thomas Broome blowing his whistle to signal an advance on the German front line at the Battle of the Somme in July 1916. He only made it about 45m (50 yards) into No Man's Land before a shell exploded nearby, knocking him out. He awoke several hours later in agony, stuck in a section of barbed wire, with deep shrapnel wounds to his leg, but he passed out again. When he came round the second time, the body of a German soldier lay beside him, with a pair of wire cutters attached to his belt. With dawn and certain death approaching, Thomas cut himself free and scrambled inch by inch back to the British front line. He was rushed to a field hospital, where he made a full recovery.

Thomas kept the wire cutters (made by ThyssenKrupp) that had saved his life.

After the war, he set up an ironmonger's shop in Uxbridge, Middlesex, and used the wire cutters a lot in his line of work, but over time, the blades blunted. Thomas took a gamble and wrote to the German company for help, and, amazingly, they sent him a pair of new blades. The wire cutters had since passed down to his grandson David, but had been left in a damp shed, so they were in a sorry state. I wondered about making a new pair of blades for them, but they were part of this unique story, so I couldn't bring myself to. You often have to make difficult decisions like this when you're working with treasured heirlooms on *The Repair Shop*. So I took the blades off, sharpened them, cleaned off the thick layer of rust, oiled the joints and they were good as new again. Now they're Thomas's family's most prized possession. Hearing a moving story like that and being able to help is the most rewarding part of what I do. It's a massive privilege.

Wire cutters

First World War soldiers used small and relatively quiet wire cutters to cut through enemy barbed wire and phone lines.

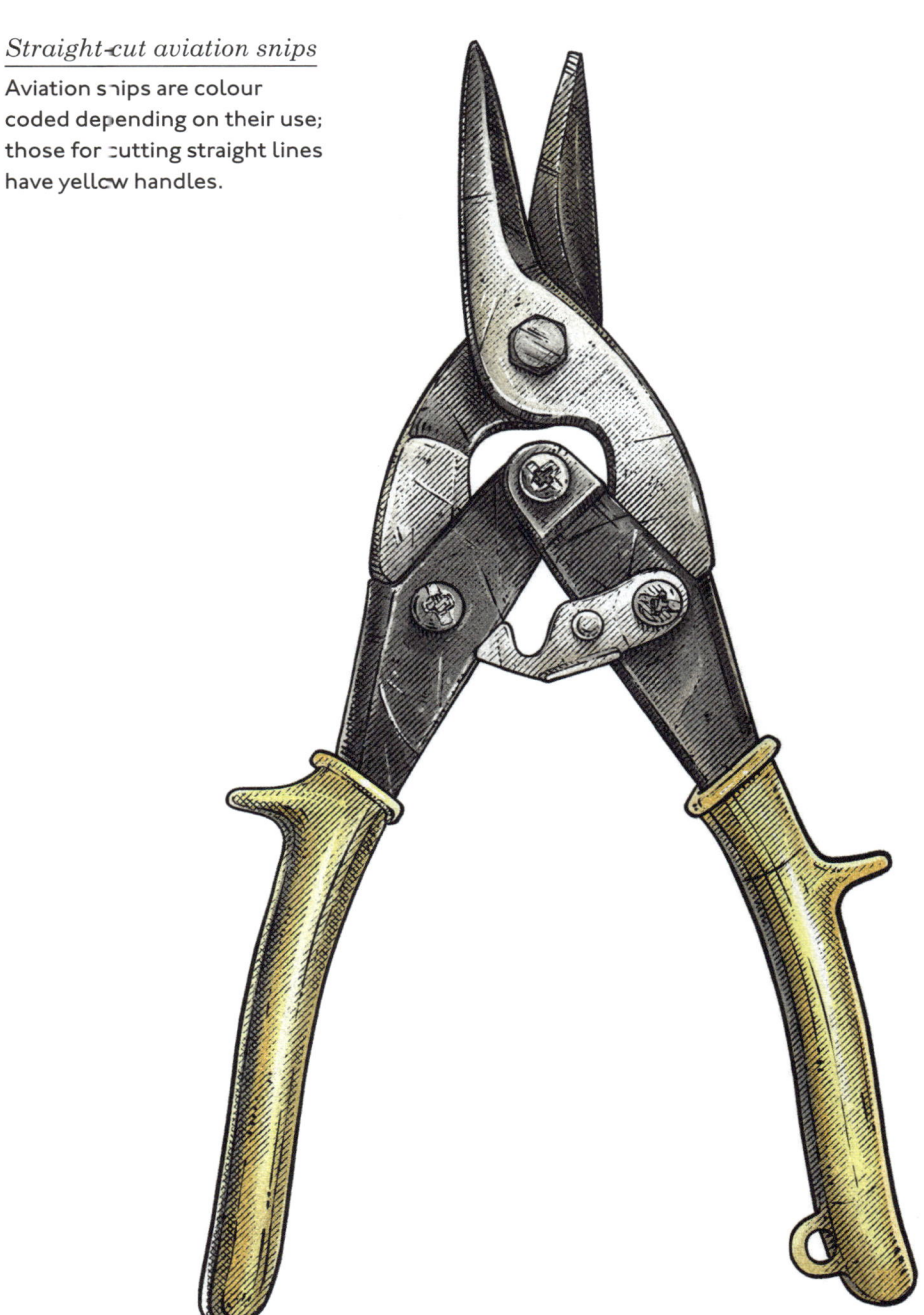

Straight-cut aviation snips

Aviation snips are colour coded depending on their use; those for cutting straight lines have yellow handles.

Tool No. **57**

AVIATION SNIPS

Aviation snips were invented and patented in 1934 by Karl Klenk, who founded a company in the USA under his own name that year. Klenk are still making the tool they're most famous for, but as part of a bigger company in Ohio. Aviation snips were designed to cut through the relatively soft sheet metal (such as aluminium) which would go on to become the fuselage (main body) of aeroplanes. Bearing in mind that the USA would go on to produce around 300,000 planes during the Second World War, these tools would have seen some serious use.

Aviation snips are also called compound-action snips, and while I know that might be the most boring alternative name ever for a tool that already has a good name, it does serve a purpose. Unlike regular tin snips, which have one pivot point, aviation snips have a second lever that increases the amount of force that the blade produces. So they're stronger than (the same size) tin snips (see page 98).

Aviation snips are designed to cut through softer metal (such as aluminium, lead, copper and zinc) as well as mild steel (steel made with a small percentage of carbon) but aren't generally made to cut through hardened, stainless or galvanized steel. They're not just used in aviation – they're used a lot by professional and amateur metalworkers in general, often to make bodywork for cars.

You can get three main types of aviation snips, and these days their handles are colour-coded so you can identify which is which. Straight-cut aviation snips are designed to cut straight lines and have yellow handles; left-cut snips have angled blades that are made to cut curves towards your left and feature red handles; and right-cut snips are made to cut curves that bend to your right and have green handles. And that's fine, as long as you remember which colour goes with which tool!

They're not just used in aviation – they're used a lot by professional and amateur metalworkers in general, often to make bodywork for cars.

Tool No. **58**

TIN SNIPS

Also called tinner snips, these snips have really wide jaws, short blades and long handles. They're heavy-duty tools, and the good ones are usually made from drop-forged carbon steel, which basically involves deforming metal by "dropping" (lowering) a mechanical hammer to shape it into a mould. Tin snips are designed to cut through sheet metal, and they work in the same way as regular scissors, with two blades joined at a pivot point that cross over each other.

Tin snips come in a few different styles and variations. First, you've got straight-cut tin snips, which are designed for straight-line cutting and can cut through soft metal, mild steel and stainless steel. Then you've got curved tin snips, which have rounded blades so they can cut curves and circles, and are best used on softer metals.

Confusingly, you've also got straight-pattern tin snips, which have got holes for your thumb and fingers, so you can keep a good grip of the tool. Duckbill-pattern tin snips have the same type of handle as straight-pattern snips but are designed for much more delicate work with their tapered blades that look like a duck's beak. You've also got bulldog-pattern snips, which as you might expect have short but very strong blades. Combined with long handles, they give you much more leverage, and therefore power, when cutting.

One company metalworkers associate with tin snips is Gilbow (The Gilbow Tool and Steel Company in Sheffield), which was taken over several times before becoming part of US tool maker Irwin (now part of Stanley Black & Decker). Gilbow started making its tin snips back in 1924, and they evolved from "tinmen's snips", which have been around since the 19th century.

In the *History, Gazetteer & Directory of Warwickshire* (just a bit of light reading) there were a few tool manufacturers, such as William Jackson from Birmingham and Thomas Matkin from Sheffield, who were advertising tinmen's snips from at least 1849.

Tin snips are designed to cut through sheet metal, and they work in the same way as regular scissors.

Straight-cut tin snips

Designed for cutting straight
lines, these snips have
straight blades and can cut
through stainless steel.

Tool No. **59**

TRADITIONAL HANDSAW

Say "saw", and this tends to be what most people think of – a wood saw with an angled back, a slightly wobbly blade and a shaped wooden handle secured to the blade with screws. The design of the traditional handsaw has remained very similar for thousands of years, and it's a classic for a reason. But let's go back to the beginning of saws.

The first true saws would have been made from smelted and cast copper and would have cut on the pull stroke (when you're pulling the saw back towards you), unlike modern handsaws, which cut on the push stroke (when you're pushing the saw away from you). A big development happened in the 1st century BCE with Pliny the Elder, the Roman author and naturalist, who worked out that if you bent the teeth slightly away from the plane of the blade, with one tooth one side of the plane, then the next the other side, you'd end up with two major advantages. Firstly, the saw cut (known as the kerf) would be wider than the thickness of the blade, and secondly, you'd end up with much less sawdust in your face. Genius. You also end up generating less friction, so the blade doesn't get as hot.

Traditional handsaws either have crosscut teeth or rip teeth, which do slightly different things. Crosscut teeth are designed to cut wood across the grain, and to do this, their teeth are angled on their inside edges, so they look like the blade of a knife. They basically sever the fibres by scoring one side then the other, leaving fine particles of sawdust. Rip teeth cut wood with the grain and push into it like a series of little woodworking chisels, leaving shavings or fibre strands behind.

One thing that makes a big difference is the number of teeth per inch on the blade, which you'll see shortened to TPI.

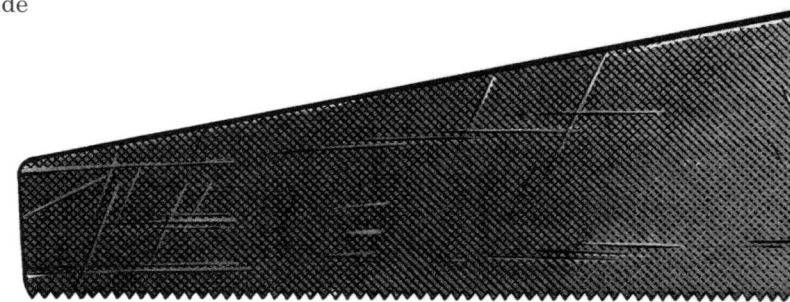

Traditional handsaw

It's thought that handsaws were used by ancient Egyptians as far back as 1500 BCE.

Simply put, the more teeth you've got, the smoother the cut will be. With rip cutting, you'll need fewer teeth than for crosscutting because the latter is harder work, so you tend to see blades with between 6 and 8 TPI. With crosscutting, it'll be more like 8 to 12 TPI. As a general rule, though, whenever you're cutting with a saw, you should aim to have two or more teeth in contact with the workpiece.

These days, though, some modern handsaws have combination saw blades so they can cut both across and along the grain. The Japanese *ryoba nokogiri* saw also allows you to do both, but in a slightly different way. It's got two rows of teeth, one on either side of the blade, and a handle usually wrapped in split bamboo. Commonly known just as the *ryoba*, this saw has been around since the end of the 19th century, and it's still a hugely popular tool in Japan.

The other major thing that makes a big difference is how sharp your saw blade is. What you want is a blade with sharp, pointy corners. There's no hard and fast rule for how often you need to sharpen your saw blade, though. It depends how often you're using it and what type of wood you're using it for. But if, when you have a good look at the teeth, they're looking a bit more on the Toblerone side of things, you can sharpen them. Get yourself a triangular saw file and a clamp, and remember that the cutting blade is on alternate sides for crosscut teeth.

The last thing to mention is that modern hardpoint saws have heat-treated teeth, which is useful because they stay sharper for longer, but when they lose their sharpness, there's no resharpening them. So it's swings and roundabouts.

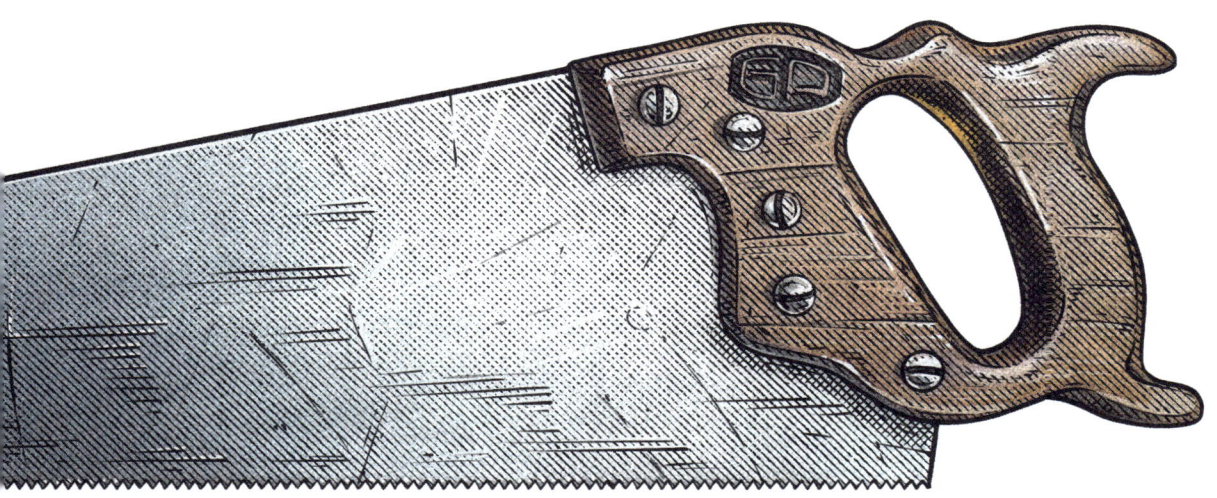

COPING SAW

Here's a tool that might transport you back to your first woodworking class. The coping saw is a really simple type of bow saw with a thin, finely serrated blade that's held under tension between pins within a C-shaped or bow-shaped metal frame. It's designed so you can cut external shapes and curves but is also really helpful for making interior cuts. In case you're still none the wiser about what coping is, it's a woodworking technique where you shape the end of one piece of wood so it neatly fits the mouldings, decorations and contours of another piece that will butt up against it. Imagine adding one bit of Victorian skirting at a right-angle to another bit and you get the idea.

You can create curves by turning the handle but you can also adjust the angle of the blade by untightening the two pins to cut even sharper curves. You can also flip the blade so it can cut on either the pull stroke (with the serrated teeth pointing towards the handle) or push stroke (with the teeth facing away from the handle), but most people seem to opt for the pull stroke.

It seems as though the coping saw descended from 16th century saws used for marquetry (creating decorative patterns on thin sheets of wood that are applied to furniture). There's a strong possibility the Romans were using similar tools – they were definitely using frame saws as there's a painting of one in Herculaneum, which, like Pompeii, was buried under ash after the eruption of Mount Vesuvius in 79 CE – but the earliest engraving of something that looks a lot like a coping saw appeared in a book by French writer André Félibien in 1676. It's a bow saw with a slim blade and serrated teeth but he labelled it "small marquetry saw". It wasn't until the early 20th century that the name "coping saw" stuck.

Coping is a woodworking technique where you shape the end of one piece of wood so it neatly fits the mouldings, decorations and contours of another piece.

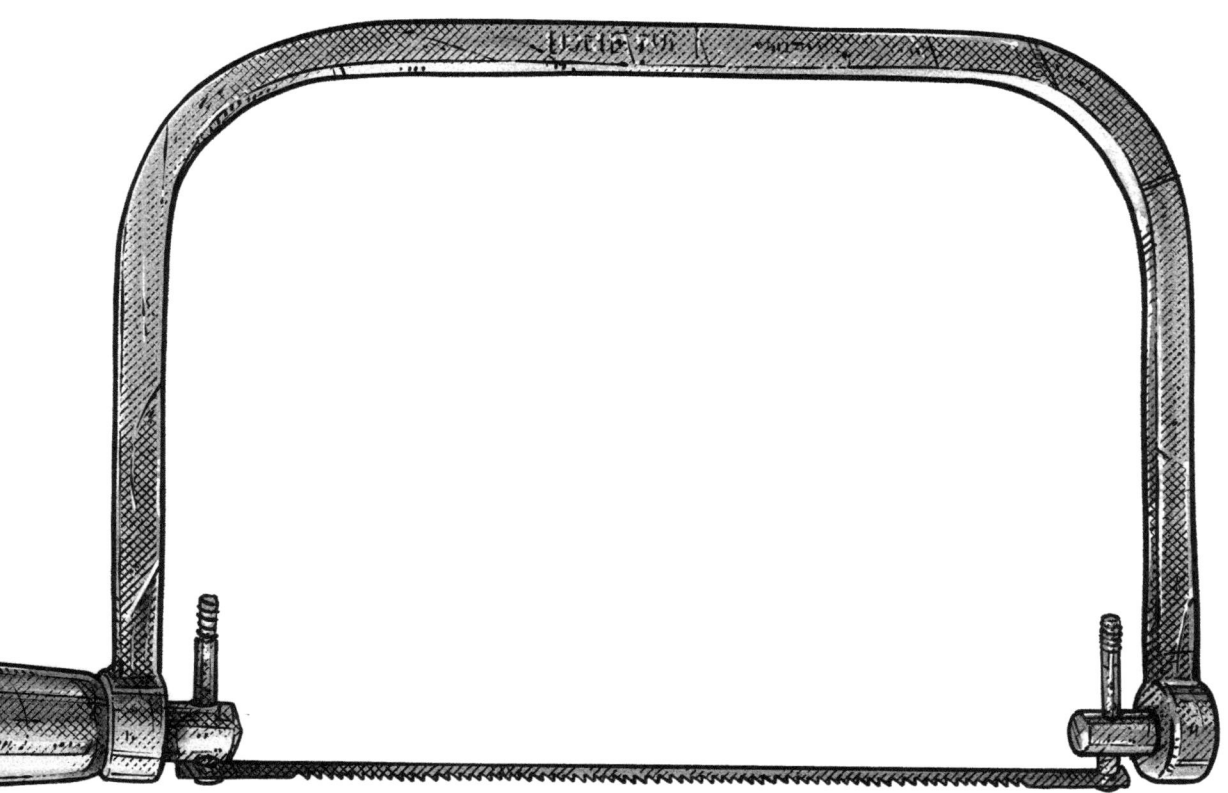

HACKSAW & JUNIOR HACKSAW

One of the first things you see hanging up in your grandad's shed is the humble hacksaw. It's a handsaw mainly used to cut metal. They usually have a C-shaped metal frame, a D-shaped "closed pistol grip" handle and a removable blade held between two pins.

Hacksaw blades come in two lengths: 300mm (12in) and 250mm (10in), but the 300mm (12in) one is the most common. You can get a number of different blades for hacksaws, featuring between 14 and 32 teeth per inch (TPI). Fewer TPI blades are suitable for softer metal, such as aluminium, and plastics, while 24 TPI will enable you to cut through steel plate up to around 6mm (¼in) thick. Although you can reverse the orientation of the blade, hacksaws are really designed to cut on the push stroke, so the teeth should point away from the handle.

The first hacksaw blades were made of carbon steel, but almost all of them now are made of high-speed steel (alloys created by mixing carbon steel alloyed with other metals such as tungsten or molybdenum), which means they're tougher and cut quicker.

The origin for the name of this saw is a bit mysterious, but it's not thought to originate with the word "hack", meaning "roughly cut" because, well, you can actually do some quite delicate work with a hacksaw.

A junior hacksaw is not a toy saw for toddlers but a smaller version of a regular hacksaw with a half-size blade, and it's mainly used for cutting through metal and plastic tubes and pipes. Like its bigger brother, you can remove the blade, but the handle is usually much simpler – just a loop of metal – although you can get fancier ones that protect your knuckles.

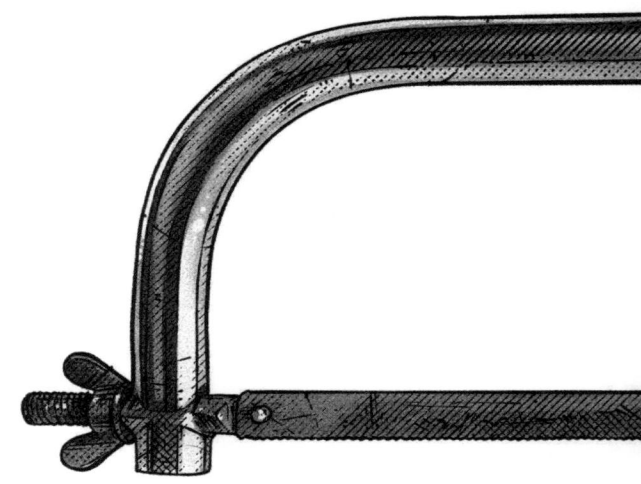

Junior hacksaw

Junior hacksaws are half the size of regular hacksaws and have a simple metal loop for the handle.

Hacksaw

Full-size hacksaws have a C-shaped metal frame and a D-shaped "closed pistol grip" handle.

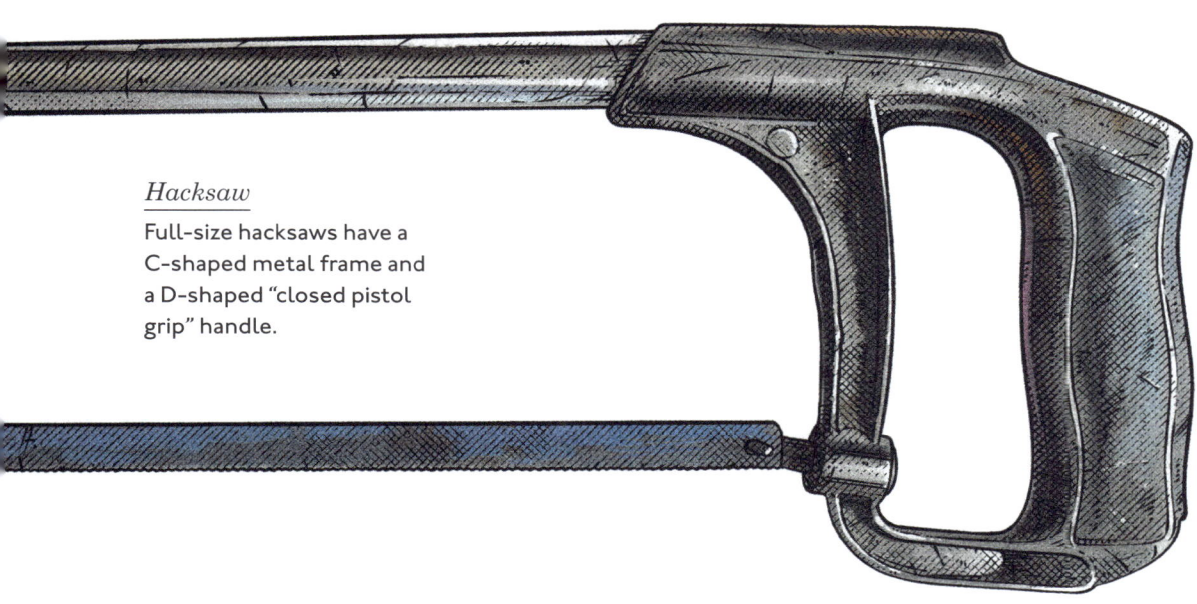

Tool No. **63**

DOVETAIL SAW

A dovetail saw is a type of backsaw, a short handsaw reinforced with a metal rib (usually iron or brass) on its upper edge. It's designed for fine joinery work and specifically for making dovetail joints, where two pieces of wood are cut to fit together perfectly, a bit like interlocking fingers (or feathers). When they're done right, these joints look beautiful, and the chances are, if you looked inside the drawers of an antique cupboard, you'd find immaculately crafted dovetail joints.

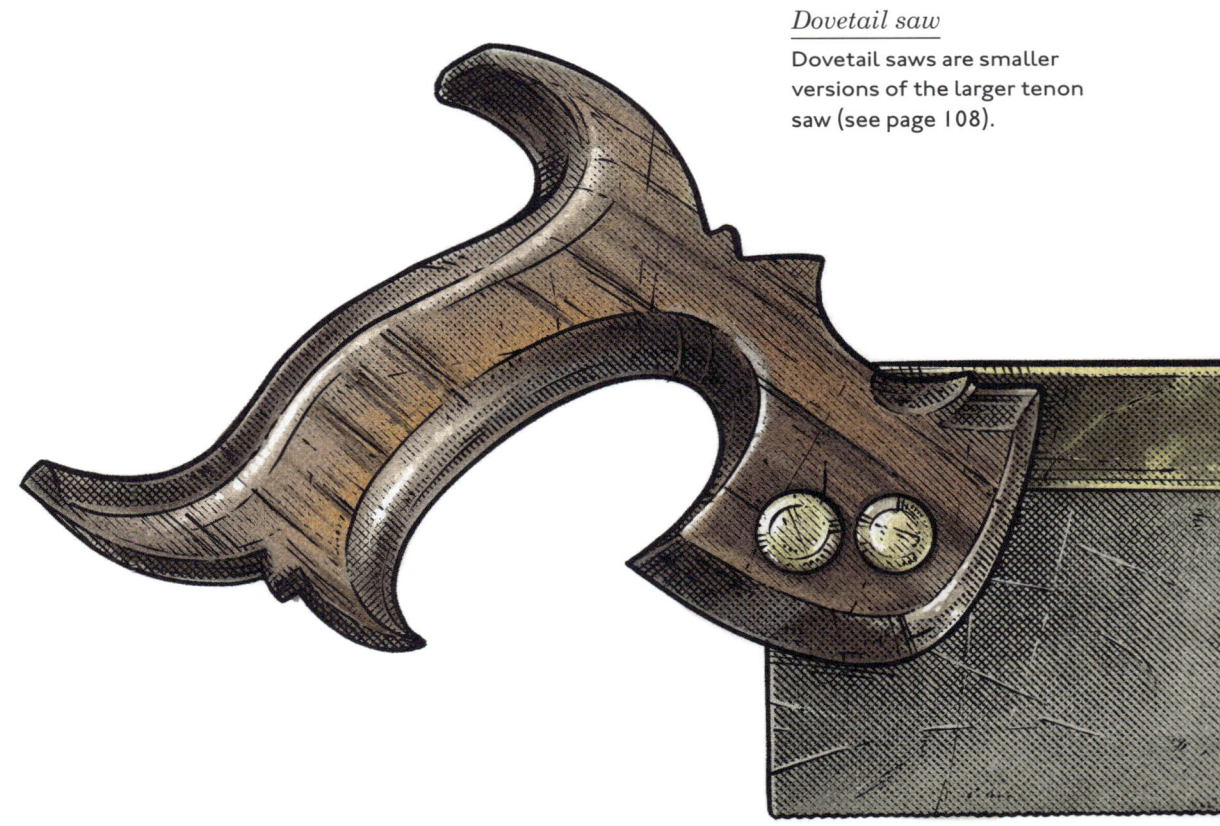

Dovetail saw

Dovetail saws are smaller versions of the larger tenon saw (see page 108).

This type of joint has been around for at least 4,000 years, and we know that because perfectly put-together dovetail joints appear on ancient Egyptian wooden coffins dating back to 1900 BCE.

A dovetail saw is between 20 and 25.5cm (8 and 10in) long and usually has between 15 and 20 teeth per inch. The shape and size of the blade has actually changed very little over the centuries, but the style of the handle has changed quite a bit, with examples from the 18th and 19th centuries looking like finely carved furniture. One of the earliest surviving dovetail saws was made by Robert or William Dalaway, Birmingham sawmakers who founded their business around 1750. It has an intricate, open "fishtail" handle, which was clearly popular as it's the style that appears in Smith's Key from 1816, one of the earliest illustrated catalogues of tools made in Sheffield. By that time, Sheffield was well established as the world centre for handsaws, thanks to its plentiful steel and the water power needed to make it.

Among the earlier Sheffield sawmakers were Kenyon, Kenyon, Plummer and Jones, who set up shop as sawmakers there in 1757. Spear & Love was formed soon after in 1760, which eventually became Spear & Jackson, who are still around today. They were followed in the 19th century by brothers Thomas and William Tyzack, who started their own sawmaking company in 1817. Their father was a toolmaker but specialized in producing scythes, so his two boys branched out, presumably buying a sawplate from the nearby rolling mill to manufacture saws and other tools made from thin strips of steel. Drabble and Sanderson were also on the scene by 1825, although they'd been working separately as individual sawmakers for some time before then.

On the other side of the Atlantic, a young Englishman by the name of Henry Disston (who'd emigrated from Derby to Philadelphia in 1833) began his own saw-making business in 1840. By 1850, he started a company that would become the biggest sawmaker in the world: Keystone Saw Works. Soon after that, he'd even started making his own steel rather than rely on importing it from Sheffield, which is what all his competitors were doing.

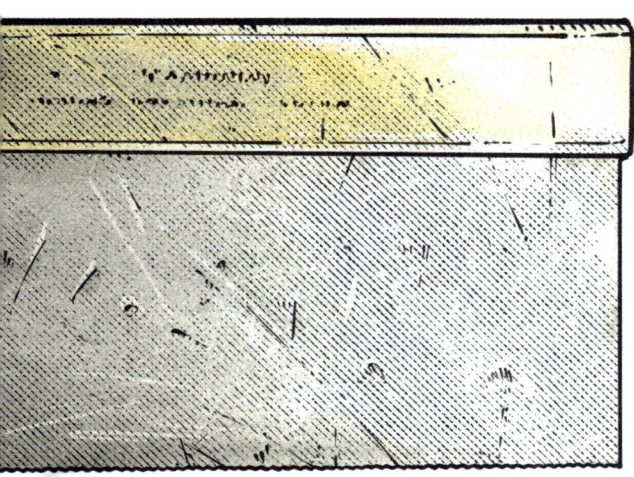

Tool No. **64**

TENON SAW

Like the dovetail saw (see page 106), the tenon saw is also named after the type of woodwork joint it was designed to cut: the mortice and tenon joint. It's a simple idea – the end of one piece of wood is shaped into a kind of peg that fits into a hole cut into the other. If you thought the dovetail joint had some history, this one goes back another couple of thousand more, because the first known mortice and tenon joint (a tusked mortice and tenon joint that uses a wedge shape to hold it together) is on a wooden structure used to reinforce a well in Leipzig, Germany, dating from around 5,000 BCE. It was a bit of a miracle that it survived intact for 7,000 years, but it was buried 6m (20ft) below the surface in a wet, cold and oxygen-free setting, so the perfect place for it. It was a huge discovery because it was the oldest intact wooden structure ever discovered, and made us realize that there was some sophisticated craftwork going on in the Stone Age.

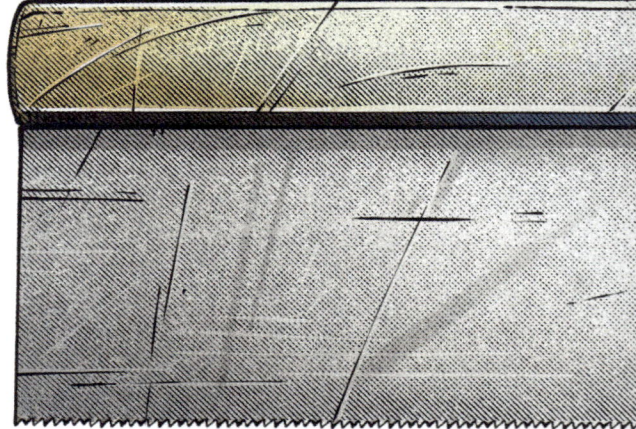

Tenon saw

Tenon saws have larger blades than dovetail saws and closed rather than open handles.

The tenon saw also appears on the same page as the dovetail saw in Smith's Key from 1816 (see page 107), but with a closed handle and a much larger blade of between 40 and 48cm (16 and 19in). As for when the first tenon saw was made, it's hard to pinpoint exactly. It's possible that Robert and William Dalaway were making them in Birmingham in the 18th century. But before then, the famous White family of sawmakers (the first sawmakers that

anything's really known about), just outside the city of London, were making saws between 1650 and 1750, so it could well have been them. The trouble is, only two of their saws have survived, and neither of them are tenon saws (although they're both backsaws). Both of those saws have ended up in America, which perhaps isn't a surprise, given that White's saws were so well regarded they were even exported to the USA back then.

Tool No. **65**

FRETSAW

A classic and unmistakeable woodworking tool, the fretsaw has a much deeper frame than a coping saw (see page 102) and a much smaller and thinner blade. It's designed for finer, more precise work than a coping saw, such as cutting fine corners, although because the blades are more fragile, they do have a habit of breaking. Unlike a coping saw, the blade is clamped in place with wingnuts rather than secured with pins.

The "fret" in fretsaw derives from the tool's use in fretwork, a type of decorative design involving metal and wood, and often featuring geometric shapes. "Fret" comes from the Old French word *frete,* which means "interlaced work". The word originated in the 14th century, but fretwork is actually far older, dating back to at least ancient Egypt, where gilt fretwork was applied to furniture such as chests and boxes.

The fretsaw was originally called a Buhl saw, which comes from a German variation for the surname Boulle. André-Charles Boulle (1642–1732) was a legendary French cabinetmaker who was renowned for veneering furniture inlaid with tortoiseshell and brass. He was so good at it that it became known as Boulle work, and examples of his craftsmanship can be seen in the collections at the Louvre, the Metropolitan Museum of Art, the Victoria and Albert Museum and Windsor Castle.

It wasn't until the 19th century that the Buhl saw became known as the fretsaw, and it was used much more widely from the beginning of the 20th century onwards.

Tool No. **66**

JEWELLER'S SAW

Also known as a piercing saw, a jeweller's saw is similar to a fretsaw (opposite) but the frame isn't as deep.

....

They are unsurprisingly used mostly by jewellers, but also by other metalworkers and hobbyists working with softer metals. Piercing saws are commonly used on V-boards (also called a birdsmouth board), a wooden board with a V-shaped cutout. The far side of the board is clamped to your workbench, while the cutout extends over the end of the workbench so you can pass the sawblade through the piece you're working on.

The blades vary in thickness depending on the type of metal you're sawing, but you can get ultra-thin blades for very delicate work. You can also get specialist spiral-shaped blades for things such as wax carving. Finer blades have more teeth per centimetre/inch than heavier blades, but because they're delicate, they can break easily. To help with this problem, basic jeweller's saws have frames that are adjustable, so they can fit broken blades. In the long term, though, most jewellers invest in a more expensive, higher-quality saw.

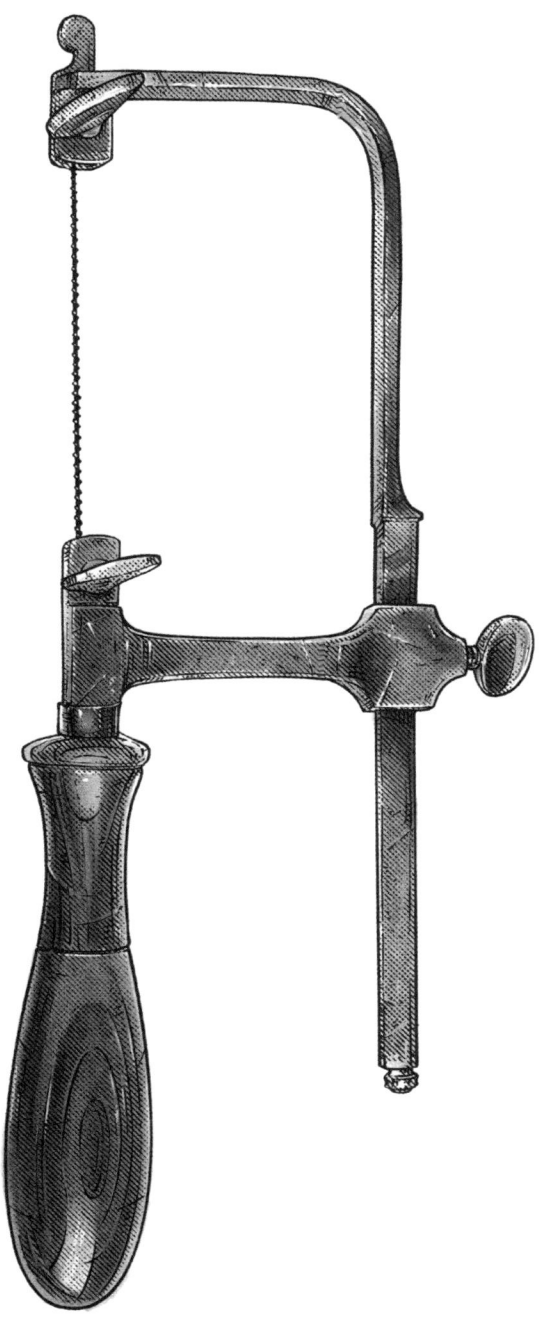

GENT'S SAW

A gent's saw is like a smaller version of a dovetail saw, only with finer teeth (up to 30 teeth per inch) and a straight inline handle. The blade length varies from around 7.5 to 25.5cm (3 to 10in) and it's used for making small joints and for precise cuts suitable for delicate furniture, musical instruments, models and toys.

A gent's saw is a beautiful, elegant tool, and the name comes from its intended use for "gentlemen". I'm assuming this was some sort of marketing gimmick designed to appeal to those folk with some cash in the bank, so your gentleman hobbyist could happily separate himself from those horrible professional tradespeople who do this sort of thing for a living. Or maybe you could only buy one if you went into the shop wearing a shirt and tie!

In Smith's Key, the famous 1816 illustrated catalogue of the tools and cutlery made in Sheffield (see page 107), there is a gentleman's saw, but it's got an angled blade between 25.5 and 51cm (10 and 20in) long, and a curved, closed handle. It looks like a smaller version of a traditional handsaw (see page 100), basically. So it's possible a gentleman's saw was a broader term used to describe a handful of handsaws used by your hobbyist as part of a sort of "gentleman's toolkit". Fancy!

Exactly when the name stuck for the gent's saw that's still made today isn't clear, but we know that William Marples & Sons (see page 71) produced a few versions, including one with a brass back and a posh rosewood handle, but that one seems to have only been produced until the late 1910s. Instead of "gent's saws", Marples called them "fancy backsaws", which tells me they had a good sense of humour about Brits and class.

It's possible a "gentleman's saw" was a broader term used to describe a handful of handsaws used by your hobbyist as part of a sort of gentleman's toolkit.

Gent's saw

Designed for the hobbyist rather than the tradesperson, gent's saws can be useful for precision work and small dovetail joints.

KEYHOLE SAW & COMPASS SAW

These two saws are designed for cutting curves (known as compasses, which is where the compass saw gets it name) in wood, softer metals and plastic. They both have long, slim, tapered blades ending in a sharp point, and either a traditional open handle (for older examples) or a pistol-grip handle, although some keyhole saws have simpler pad handles that follow the direction of the blade.

Both saws look very similar, but the compass saw has a larger, longer blade (up to around 46cm/18in compared to around 25.5cm/10in for the keyhole saw) and is made to cut through thicker materials. Both of them are really useful in tight spaces where other saws, with their wider blades or frames, can't reach. Although both saws have a pointed tip, they weren't actually designed to make a hole so you have an edge to work off when cutting interior shapes in wood. But that is what they get used for a lot. And I can't pretend that I haven't used my keyhole saw to cut rectangular and square holes in plasterboard for plug sockets and light switches. Both saws do have a history of not being used exactly for the purpose they were intended. In Chas A. Strelinger & Co's *Wood Workers' Tools* catalogue from 1895, they're very upset at how the keyhole saw gets used: "It's used in turn to cross-cut, miter and rip, and as a rule it is the most thoroughly abused and maltreated tool in the carpenter's kit." On one hand, while that might be a bit of a shame, I

know full-well that when you pick up a tool you fall in love with, it's sometimes hard to put it down!

Both are really useful in tight spaces where other saws, with their wider blades or frames, can't reach.

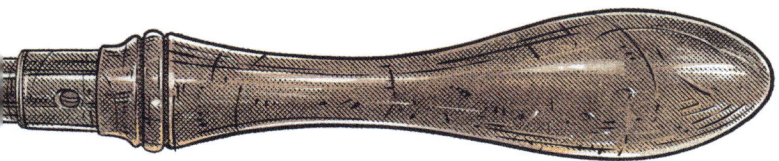

Keyhole saw

Keyhole saws have shorter blades and often simple pad handles.

Compass saw

Compass saws have longer blades suited to cutting thicker materials.

AXES

Axes are among our oldest tools, most likely dating back at least 1.6 million years. At that point, they were made of flint, with a sharp cutting edge at one end that tapered down to a rounded surface at the other to fit in the palm of the hand. These were probably used for all sorts of different tasks, from butchering meat to chopping wood and digging up plant tubers. Exactly when a handle first appeared on an axe is tricky to work out, but archaeologists in Australia did find a fragment of basalt that had been ground and polished in a way that didn't look like it was made to be held in a hand, and that dates back around 49,000 years. There are many different types, shapes and sizes of axe, from double-sided felling axes to lightweight hatchets. And then there are tools closely related to axes, such as the ancient adze.

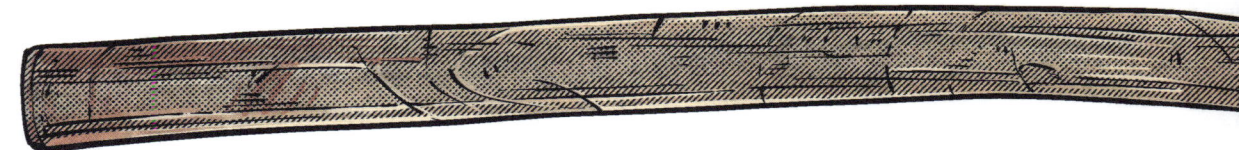

ADZE

Thought to date back to the Stone Age, an adze is like an axe but it has an arched head with a blade set at a right angle to the handle.

There are two main types of adze – one has a short handle and can be used with one hand, while a bigger-bladed, longer-handled form (known as a foot adze) is designed for more heavy-duty work with both hands. Both versions are used for shaping, carving and smoothing wood.

Some types of foot adze – railroad adzes, for example, can weigh up to around 2.2kg (48lb) – were used for shaping railway sleepers. Adze is a funny old word and dates back to at least 900 CE. It's thought to originate from the Old English word *adesa*, although no one knows where that word came from. It may have been the previous name for the tool, though, in England anyway.

It's clear that the adze was one of the go-to tools for carpenters plying their trade in ancient Egypt. It seems to have occupied a special place in Egyptian culture because the adze, shown next to a block of wood is actually an Egyptian hieroglyph, which symbolizes the word "chosen". For that reason, the hieroglyph is almost always found carved in the tombs of Egyptian pharaohs to describe that they've been chosen by a particular god. In the Tomb of Ani (around 1250 BCE), the contents of which are in the British Museum, a whole collection of carpenter's tools was found, including a heavy adze made of bronze, strapped to a wooden handle with leather lashing.

Adzes have also been found in New Zealand, used by the ancestors of the Māori for making canoes. Many of them were made from basalt and greywacke, a type of hard sandstone. They've also been excavated in Alaska, with blades made of nephrite, also called jade. Adzes are still used by people in Indonesia and Papua New Guinea in ceremonial craft-making and by traditional craftspeople, such as coopers, in many parts of the world.

Tool No. 71

SPLITTING AXE

This is your classic axe for chopping wood, and it's well suited to that task with its heavy axe-head and hardened poll (the back end of the head), compared to your lighter, narrower felling axe.

For splitting wood, you want something a bit thick, so you can deliver more force to split apart the wood fibres. A heavier head also has the major benefit of minimizing the "axe head stuck in the wood" situation, which is annoying when you're trying to get through something repetitive quickly. Although having said that, a lot of folk love splitting wood!

Estwing is well-known for the quality of its hammers (I've got three of them and they're lovely things), but it also makes fine axes and specialist geological tools too. Neil Armstrong used an Estwing geologist's pick axe when he was training for the mission to the Moon. The company's origins go back to 1923 and the town of Rockford in the far north of the state of Illinois. It was set up by a young Swedish immigrant called Ernest Estwing, who'd moved to the USA in 1900. He bought himself a set of books about the machinist's trade and managed to both learn English and get himself a profession. Talk about efficiency! While he was working as a contractor machinist, he noticed that a lot of workers were suffering from joint problems after swinging wooden-handled tools all day, so he came up with a solution in his basement (like all good American inventors).

He reckoned that drop forging axes, hammers and hatchets from a single piece of steel was a much better idea for making the tool stronger, lighter and better balanced. But even more than that, it would basically render the tool pretty much unbreakable. Over 100 years later, the company is still built on these principles, and it's served them well. Everything's made in the USA, they're still in Rockford and they produce some of the best splitting axes around. They also make splitting mauls, which look like a cross between an axe and a sledgehammer. They get through wood with brute force, and are helpful for getting through those bigger, heavier and trickier bits of wood.

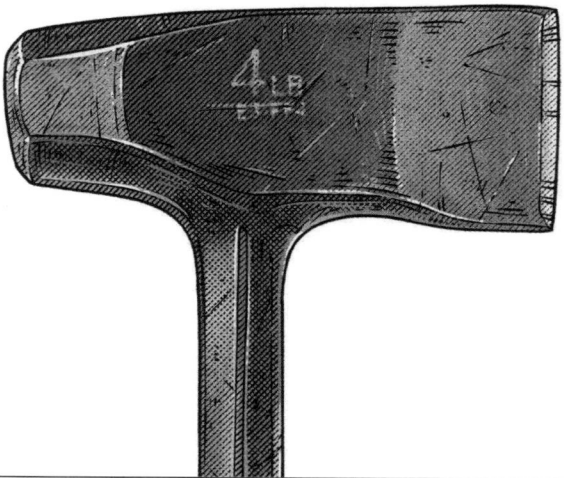

Tool No. **72**

DOUBLE-BIT AXE

So called because they have two cutting edges (or bits) on the axe head, these axes are really useful because if you're doing one type of work with them and the blade starts to dull, you can just flip it round and use the other cutting edge. Or if you're doing a couple of tasks with them, you can use one edge for the messier work (say, where the axe head may come in contact with the ground when you're chopping tree roots), while you use the other for tasks such as chopping trunks.

One of the most renowned double-bit axe-makers was Collins & Company in Connecticut, USA. Founded in 1826 by brothers 24-year-old Samuel Collins and 21-year-old David Collins, they specialized at first in just making axes. To be fair, they didn't need to diversify, because at that time everyone, from loggers to builders, was after a good axe. And maybe more importantly, until Collins & Company started making axes, you couldn't just buy a ready-to-use axe. You'd buy an axe head from a smith, but then someone needed to sharpen it and fit a handle to it. So Collins & Company not only produced high-quality axes, it also provided the solution to a problem. Demand soon went through the roof and the company began making other tools, such as adzes picks and sledgehammers. In 1859, it was producing 2,000 tools a day, and by the time Samuel died, in 1871, the company had manufactured around 15 million axes and was selling $1 million worth of axes each year.

CHAPTER 5
CARVING

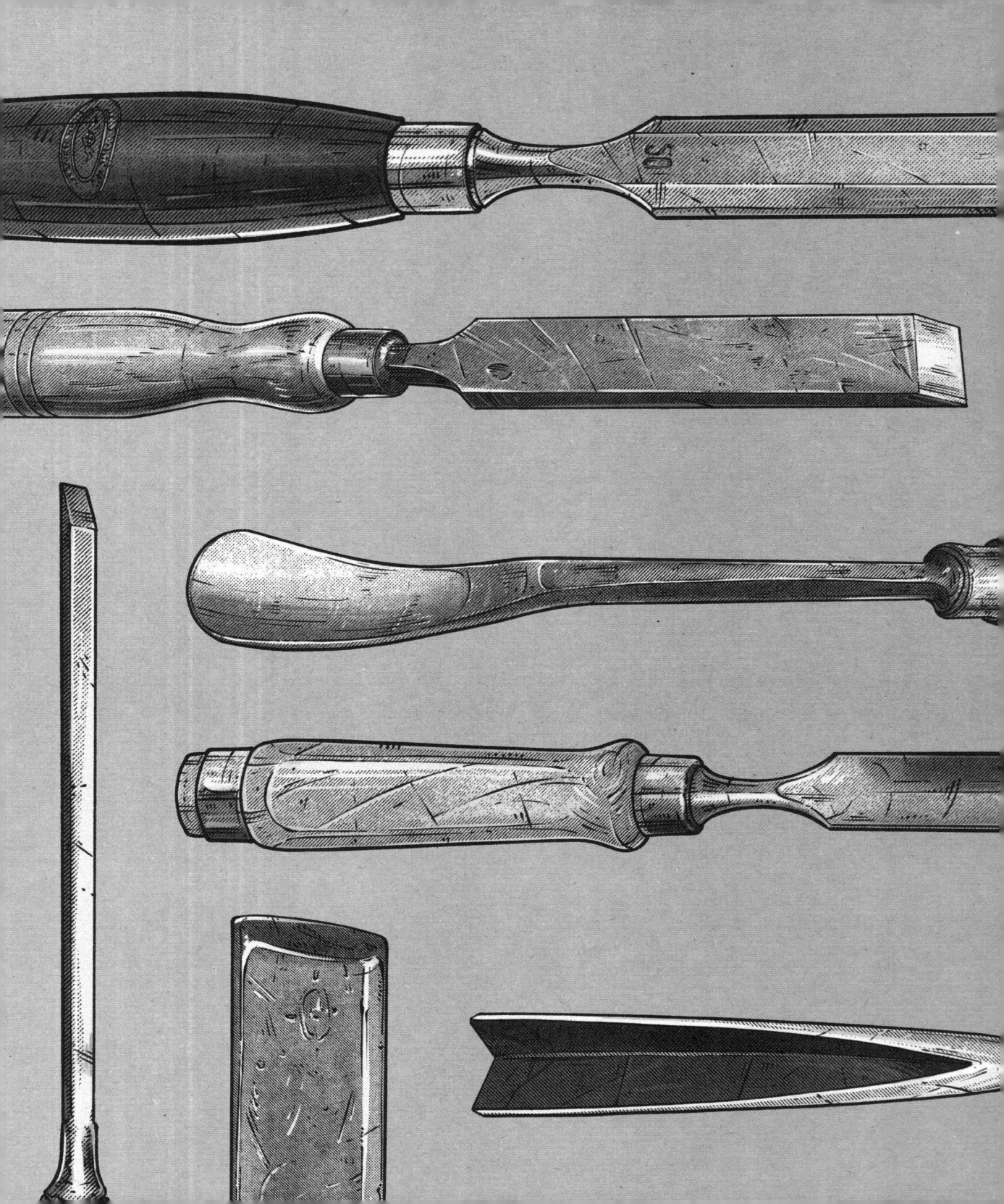

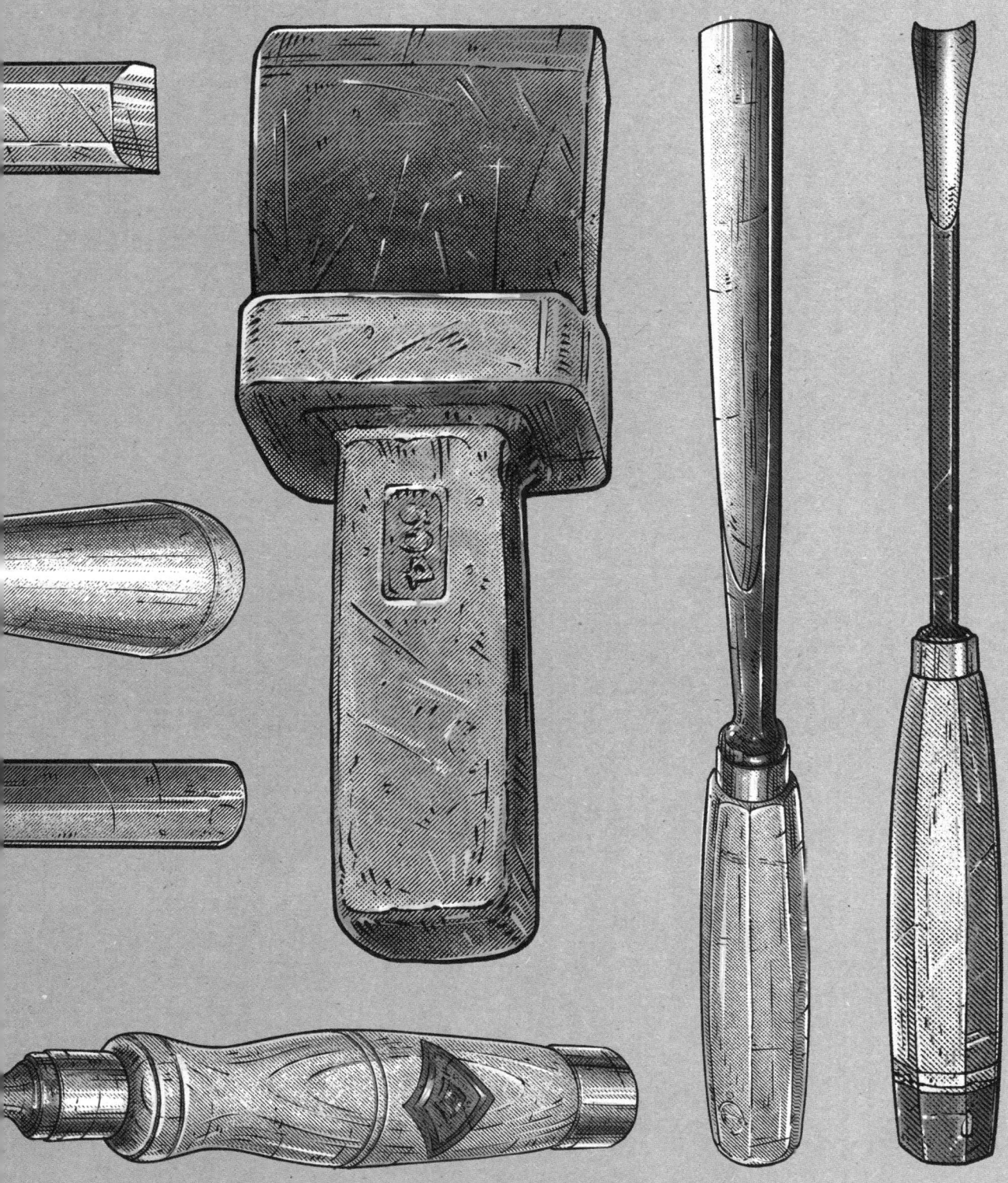

CHISELS

A chisel is a tool with a cutting edge that you drive into wood (but also stone and metal) using either your hand, a hammer or mallet, or with mechanical help. The first tools that resembled modern chisels were made of flint and date back over 10,000 years. But flint's not the best choice for a chisel because it fractures, so it was replaced by more durable ground and polished rocks. Then around 5,000 years ago, after our ancestors learned to extract metal by heating up rocks, the first copper chisels were made, followed by bronze chisels (created by mixing copper and tin). The Romans then used iron chisels, and these really wouldn't have looked very different to the chisels used by carpenters today.

Mortice chisel

Joseph Moxon was the first person to write a DIY book in the English language, and he was a big fan of the mortice chisel. Get hold of a copy if you can.

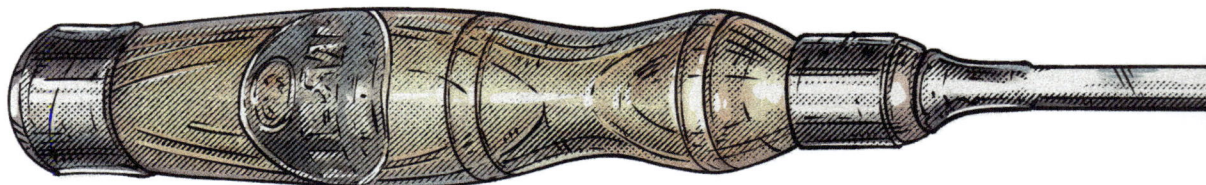

Tool No. **73**

MORTICE CHISEL

A mortice chisel is a square-edged chisel that's used mainly to cut out a mortice, and in case that hasn't helped explain it, a mortice is a rectangular or square hole. It's one half of a mortice and tenon joint (the tenon being a projection in a bit of wood that fits into the mortice).

You use a mortice chisel by striking the end of the handle with a mallet or hammer to chop out pieces of wood and lever them out. Seeing as it does take quite a bit of punishment, everything about it needs to be strong, so the blade is made of steel with a sturdy neck that connects to a strong hardwood handle so it doesn't split.

A person who's a bit of a hero in the vintage tool world and was always going to get a mention in this book is Joseph Moxon. He was the first person to write a DIY book (in the English language) and it may come as a surprise to find out that he wrote it in the 1670s. It was called *Mechanick Exercises: or, The Doctrine of Handy-Works* and in the year after the first part of the book was published, he became the first tradesman to be elected as a Fellow of the Royal Society. He broke down some serious class barriers.

The reason I'm bringing him up here is because he's got a lot of time for the humble mortice chisel and goes into great detail about cutting out a mortice. It may seem quite amazing to read an instructional guide dating from the 1600s that you can follow, but the techniques haven't really changed. What has changed is the metal used to make the blade – now carbon steel, which has been tempered and hardened. There's also usually a leather washer between the blade and the handle, which absorbs the impact of the mallet.

A mortice chisel is one of the most satisfying tools I've ever used. That feeling when you cut your first mortice and tenon joint and you realize that you've done your measurements right and everything slots together and fits (and it's not too tight so you have to hammer it in) is amazing – give it a go! It's a good one for you to try if you want to get in to woodworking, because there are no funny angles, so you can just use an engineer's square (see page 16). Once you've mastered it, you can incorporate a mortice and tenon joint into all sorts of things, such as when making shelves or a chest of drawers.

HOTCUT HARDY

You kind of know this is going to be a blacksmith's tool from the name. This chisel can have either a straight or curved cutting edge with a square shank.

You don't move the hardy from this position in the anvil; instead, you get the material you want to cut nice and hot in the forge before bringing it over to the anvil. Then you strike the material with a hammer, forcing it down onto the hardy, which cuts through it. Typically, you rotate the object after a few hammer blows. You need to keep the hotcut hardy sharp to make it effective.

A lot of blacksmiths make their own hotcut hardys because, well, they've usually got everything they need to make it – a forge, anvil, hammer, tongs and steel to start forming it.

Hotcut hardy
The cutting chisel fits in the square-shaped hardy hole of the anvil with the cutting edge facing upwards.

PARING CHISEL

Paring is just cutting away an edge or surface, and this chisel does that better than anything. It has a long, thin blade and isn't designed to be whacked with a mallet.

You can tell that from the shape and structure of the handle, which used to be made of boxwood and connected to an octagonal bolster. A paring chisel is designed for the finer work of tidying up, smoothing out and shaving the joints that you've been shaping with your firmer chisel (see page 130). They're also really useful in tight spaces.

Joseph Moxon gets quite passionate about how you should hold a paring chisel in his DIY guide from the 1670s (see page 125): "the blade is clasped upon the outside of the hindmost joints of the fore and little fingers, by the clutched inside of the third and middle fingers of the right hand." He goes on to say that, "this may be a preposterous posture to handle an iron tool in" but explains his reasons: "should

workmen hold the Blade of the Paring-Chisel in their whole Hand, they must either hold their Hand pretty near the Helve (handle), where they cannot well manage the Tool, or they must hold it pretty near the edge, where the outside of the Fingers will hide the scribed line they are to pare in." Fair enough, Joseph!

Paring chisels used to have straight edges, but at some point in the late 1800s, bevel-edged versions became available for those with some cash in the bank. The angle of a paring chisel's cutting edge is still typically between 20 and 25°. The lower the angle, the less force you need to push the chisel into the wood. So for finer work where you need as much control as possible, go for an angle of 20° or below.

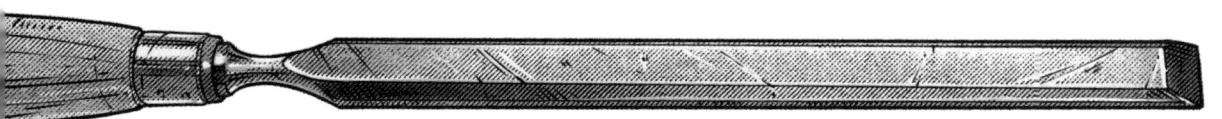

BEVEL-EDGED CHISEL

"Bevel" is one of those words that some people won't be familiar with, and to others who are familiar with the word, it can be tricky to explain it to someone else when you're put on the spot!

One day, on set at *The Repair Shop*, I was with Steve (Fletcher, our clock and watch expert), who needed a piece of glass with a 50mm (2in) bevel on it. He asked one of our fantastic assistants if he could pick one up. "No problem", the assistant answered, adding, "Just one question, though. What the hell is a bevel?!" So, here's a simpler and hopefully better answer to the one that Steve and I cobbled together at the time: a bevel is a slanted edge.

So bevel-edged chisels have a parallel blade with distinctive sloping edges. It's a broad category because there's quite a lot of variety in the thickness, length and angle of the blades. Heavy-duty bevel-edged chisels are thick, robust and can cope with being struck with a mallet, whereas finer, longer versions are designed for more delicate hand work. The bevelling means

that the chisels are good at getting into corners and working with joints with acute angles, such as dovetails. That means they get used a lot in furniture and cabinet-making.

Chisels with bevelled edges have actually been around for a long time, only they wouldn't have been known as bevel-edged chisels. There are firmer/former chisels and paring chisels with bevelled edges in Paul Hasluck's *The handyman's book of tools, materials, and processes employed in woodworking,* published in 1903. But the Dutch were clearly working with bevelled chisels since at least the 16th century, as one was found on Novaya Zemlya (an island chain in the Arctic ocean in the far northeast of Europe), which had been the setting for a disastrous mission to try and reach Asia via the Arctic in 1596.

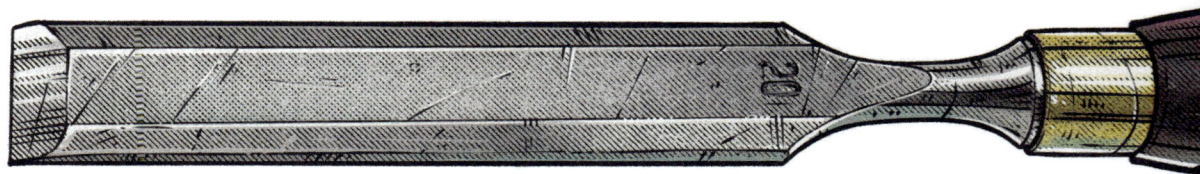

Bevelled chisels were used by the Dutch explorers Willem Barentsz and Jacob van Heemskerck in the 16ᵗʰ century.

Two famous Dutch explorers – Willem Barentsz and Jacob van Heemskerck – ended up stranded there after their ship had got stuck in ice, so they built a cabin and desperately tried to survive a polar winter with 15 members of their crew. Amazingly, 16 of the 17 survived in the cabin, although four of them weren't so lucky in an open boat on the perilous return journey to Amsterdam. Among the tools that the men abandoned was a chisel with clear bevelled edges, but it doesn't have parallel sides, like the ones we know today. It's tapered, getting wider towards the blade.

Bevel-edged chisels were the first set of chisels I bought, back when I was working as a set designer for Rankin (the British photographer, director and co-founder of *Dazed & Confused* magazine). They were actually one of the few things I bought new and they were Irwins. I don't do tons of woodwork anymore, but I did at Rankin's, and they were a really good entry-level introduction to chisels. They weren't a fancy or expensive set but they've served me well, and I've still got them in my shed at *The Repair Shop*.

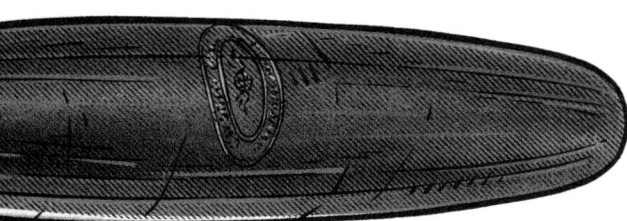

Bevel-edged chisel

Put simply, a bevel is a slanted edge, so as the name suggests, these chisels have blades with distinctive sloping edges.

FIRMER CHISEL

A firmer chisel has a thick, flat blade with parallel sides and is often used with a mallet for making deep cuts for joints.

There's no clear explanation for how the firmer chisel got its name, but the most likely explanation is that it comes from the French word *formoir* meaning "to form". I prefer that over the other suggestions that it could be something to do with the way the chisel was constructed (of solid steel), or because it's a pretty heavy-duty chisel used for removing material quickly.

We do know that "firmer" and "former" were being used interchangeably in the 18th century (and possibly some of the 17th) for the name of a straight joiner's chisel, which is something author Richard Neve

mentions in his 1736 book *The City and Country Purchaser's and Builder's Dictionary*. But if we go back a bit further, to Joseph Moxon's early DIY guide from 1677 (see page 125), there's no mention of "firmer" chisel anywhere – only "former". Moxon lists it as one of the four types of chisel along with the mortice, paring and ripping chisels (see pages 125, 127 and 218). And he's kind enough to tell us the reason for the name: "They are called formers because they are used before the paring chisel."

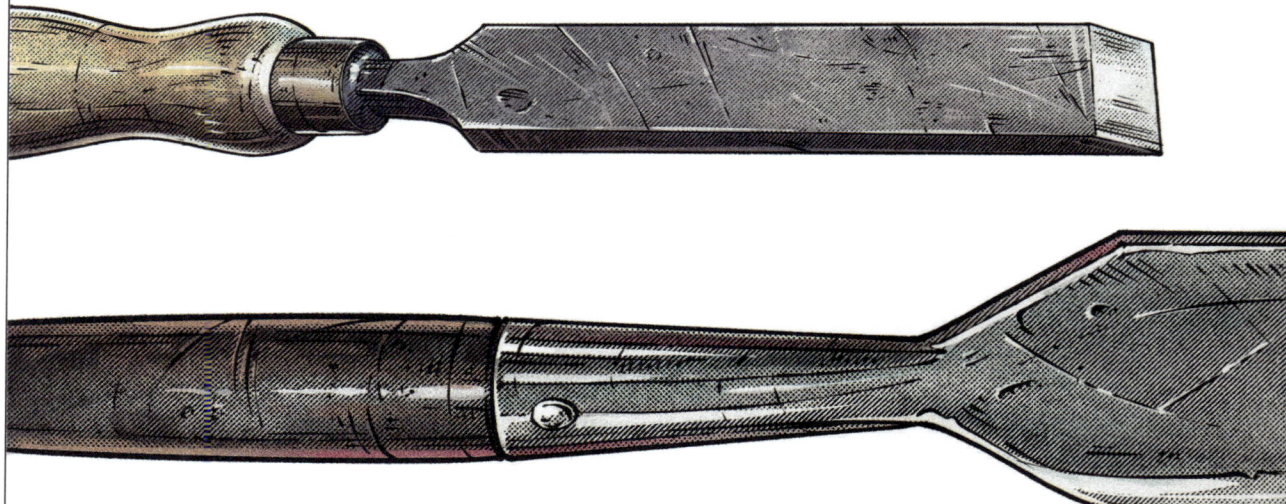

Tool No. **78**

SLICK

With its sheer size, baseball-bat-style handle and long, flat blade, a slick is a unique tool. It's designed to be used with both hands to pare long surfaces of wood with the grain.

A slick is like a really big version of a paring chisel (see page 127) and has been used for thousands of years to help craft some of the most important things in human civilization: boats and houses. Slicks made for building boats typically have a handle that ends in a round knob and were designed to be pushed with your hands, but slicks used for framing houses have shorter handles with a steel ferrule because they were designed to be struck with mallets.

Traditional house builders and wooden-boat builders still use slicks, and, as you can imagine, they become treasured possessions. You can find some beautiful old ones for sale on vintage tool sites, which aren't cheap, but when you see the size and the quality of the craftsmanship, you wouldn't expect them to be. There's a

continuing demand for slicks, so one company still producing them is Robert Sorby. It produces two versions of a timber framer's slick – both are 71cm (28in) long, but one is 60mm (2⅜in) wide and the other is 85mm (3⅜in).

Robert Sorby was established in Union Street, Sheffield, in 1828. At some point in the middle of the 19th century, the company started using a kangaroo motif, and it seems that was to capitalize on the emerging markets in British colonies. They admit that was the reason for introducing the Kangaroo Brand around the 1880s on their tools. Robert Sorby moved a few hundred yards to another site in Sheffield in the 1890s, named Kangaroo Works. A stone kangaroo stood proudly over the entrance to the building. The building is no more, but the arch and kangaroo were saved (hooray!) and now form part of the Hawley Collection, which is housed in the Kelham Island Museum, in Sheffield's city centre. It's one of the most incredible collections of tools in existence anywhere in the world.

GOUGES

Gouges are basically chisels with concave or curved blades and have been around for approximately 9,000 years. They're used for all sorts of things, from hollowing out furniture and cutlery to carving beautiful letters into panels. Some of the best gouges were made in London and Sheffield in the 19th century, and one company name occupies a special place in the hearts of amateur woodcarvers: Addis. And for good reason – the Addises made some of the finest and even now most highly sought-after gouges ever made, but it was not one big happy family.

Samuel B. Addis had been an edge-tool maker since at least the 1780s, based in Deptford, London. Two of his grandsons, Samuel Joseph and James Bacon, would go on to become the most famous of the Addis toolmakers. Samuel Joseph was born in 1811 and became his father's apprentice before setting up his own shop in the 1840s. His brother, James Bacon, was born in 1829, and also became an apprentice to his father. But they very much went their separate ways.

Both men separately exhibited their tools at the famous Great Exhibition in London in 1851, a massive international exposition co-organized by Queen Victoria's husband, Prince Albert. It was held in Hyde Park in a vast cast-iron and glass structure known as the Crystal Palace, and it housed around 15,000 exhibitors. Just under 3,000 of the exhibitors were awarded The Prize Medal to recognize that "a certain standard in excellence production or workmanship had been attained – utility, beauty, cheapness, adaptation to particular markets, and other areas of merit". Only James won a Prize Medal, while Samuel was honourably mentioned. Samuel complained that James had actually purchased the tools from him and then removed his name, but James claimed the reverse had happened. No one knows what actually happened, but James was either very proud of his Prize Medal or wanted to rub his brother's nose in it (or both), because he started adding the inscription "Prize medal" to his tools.

Samuel continued working in London producing tools with his S.J. Addis imprint, and company adverts from 1854 claim that he alone was the "sole inventor of the improved carver's tools exhibited at the Great Exhibition". Meanwhile, James was struggling financially, but his fortunes took a turn in 1862 with another Prize Medal, this time at the Great London Exposition. He started stamping his tools with "None Genuine Unless with the Brand – J.B. Addis Prize Medal 51 & 62", but it didn't do him much good as he went bankrupt the following year.

In 1864, James moved to Sheffield and started working for Ward & Payne, another well-known name among carvers, dating back to 1803. He was introduced to the Edge Tool Forgers' Union, but they soon got wind of the scandal after the Great Exhibition and, under pressure from the union, James was dismissed by Ward & Payne only to be rehired independently. In 1870, James won a Gold Medal at the Workmen's International Exhibition in London. The Prince of Wales even became his patron.

Samuel died in 1871, and Ward & Payne bought the rights to his business assets, including his trademark. Then in 1875, Ward & Payne rehired James, but he left after just a few months, and in a full-page advertisement taken out that year, he made it very clear that he had nothing to do with Ward & Payne. That was the year of the Philadelphia Centennial Exhibition (the first World's Fair to be held in the USA), at which both James (J.B. Addis & Sons) and Ward & Payne exhibited. Both companies were awarded Prize Medals at the exhibition, and James added it to the inscription on his tools, which now read (slightly ridiculously): "Prize medals 51, 62, 70, 71, 73 & 76".

James B. Addis died at his house in Sheffield in 1889 only months after David Ward died suddenly of a heart attack at Sheffield railway station. Unlike Ward, with his grand funeral and large fortune, James had little to show for his colourful life, and his death wasn't even reported in the press. His business was continued by his son and then his widow, before eventually being liquidated in 1970. Meanwhile, Ward & Payne kept using the S.J. Addis brand name all the way through to the 1960s, until the company folded in the 1970s.

Tool Nos. **79-80**

FISHTAIL GOUGE & ALLONGEE

With a tapered blade resembling the fan-like tail of a fish, a fishtail gouge is really good for cleaning out surplus wood in corners and getting into tight spaces.

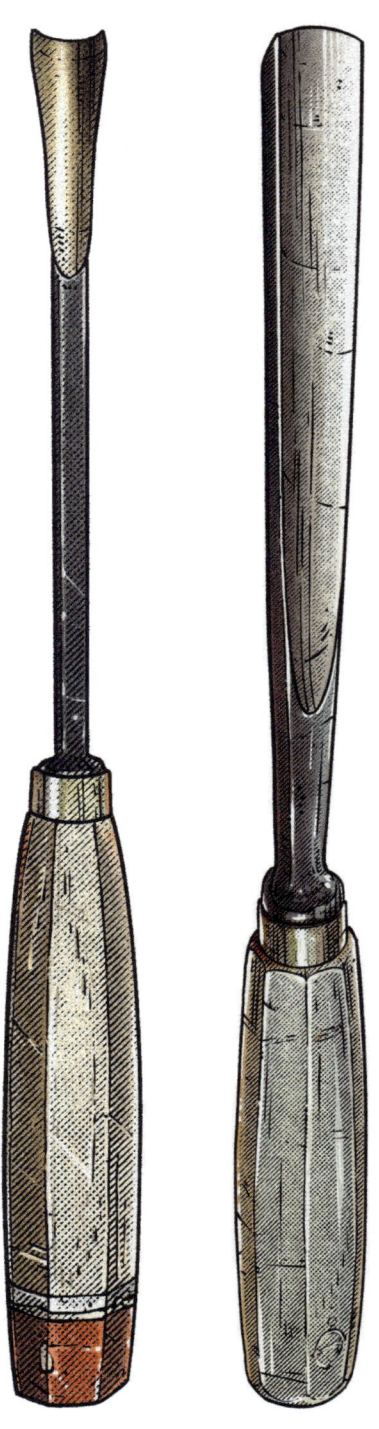

They're often used for detailed work such as carving letters and undercutting – angling cuts to remove wood under the edge of a part of a relief carving, which has the effect of creating deeper, dramatic shadows, creates the illusion of depth and also hides joint lines. An allongee is similar to a fishtail gouge but its blade tapers out all the way from the handle. They're often used to rough out an initial design.
One name that a lot of carvers seem to talk about is Pfeil (which is German for "arrow", and this symbol appears on its tools), a toolmaking company founded in 1902 in Langenthal, Switzerland, where it is still

Fishtail (left)
So called because the fan-like tapered end is said to resemble a fish's tail.

Allongee (right)
An allongee is tapered all the way up the shaft, making it a useful tool for heavier wood removal.

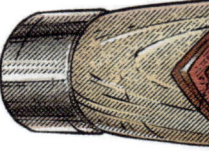

based today. The first tools it produced were surgical instruments and professional cutting instruments, but it diversified in 1942 and began supplying high-quality steel carving tools. Just bear in mind that the numbering system is different to the Sheffield List (see page 136), so takes a bit of getting used to if you're used to the Sheffield way.

The name "allongee" comes from the French *allongée*, meaning "elongated". The fishtail gouge got its name for less evocative reasons – apparently the tapered end looks a little like a fish's tail. If you ever confuse your fishtail gouge with your allongee, it might be useful to remember that the allongee is elongated all the way up the shaft, whereas the fishtail is only tapered at the "tail" end. Both are great additions to your toolkit if you're regularly needing to do detailed carving work or delicate wood removal in tight spaces.

Tool No. 81

V-SHAPED GOUGE

As the name suggests, the cutting blade of this gouge forms a V-shape, but there are several different V-shaped gouges available, with the angle of the "V" varying between 30° and 90°. The most common are 45°, 60° and 90°, but the 60° is regarded as the most versatile.

A V-shaped gouge is also known as a parting tool, because it's mainly used to separate two sections of wood. Not all V-shaped gouges have straight edges, though. You can also get a winged parting tool, which forms a shallow V with rounded sides and is used for cutting V-grooves and for outlining lettering.

V-shaped gouges can be useful for softening work you've done with other gouges by rounding the top of a carved edge. They are also used by wood carvers for adding fine details to almost-finished works.

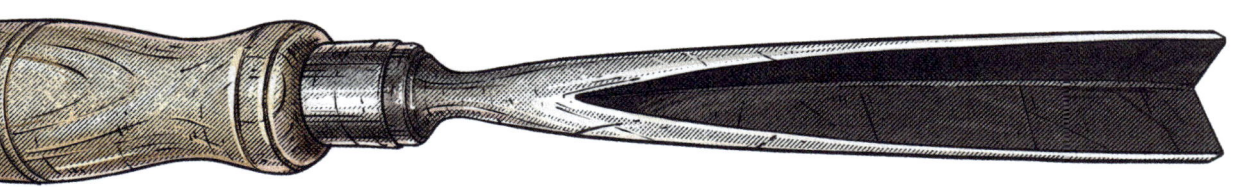

U-SHAPED GOUGE

These come in a range of different sizes, and the blade is a curve that roughly forms a U-shape. The curve of the gouge is known as the sweep.

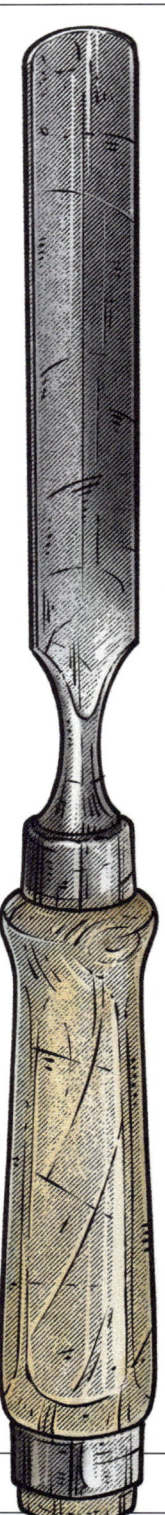

U-shaped gouge
Designed for removing sections of wood quickly and easily, U-shaped gouges are available in different widths and depths.

Back in the 1880s, a number system that became known as the Sheffield List was developed in England so that wood carvers could easily identify which carving tool does what. #1 is a straight chisel, #2 a skew chisel, and numbers 3 to 11 are gouges. #3 is the gouge with the shallowest curve and works well for levelling or flattening concave regions. Moving up from #3, the sweeps get more curved, reaching numbers 10 and 11, which are the U-shaped gouges. Basically, for gouges the rule is: the higher the number, the sharper the curve. Beyond that, you go into V-shapes (see page 135). The list is still used, although different tool producers now use different systems.

U-shaped gouges take out unneeded wood quickly and, unlike a chisel, the corners don't dig into the wood and get stuck. Deep, narrow U-shaped gouges, which are mostly used for adding fine details such as lettering and narrow lines, are called veiners. They're thought to have got this name from the way they were used to add "veins" into carved wooden leaves. Another tool name you might come across is a "fluter" – a wider U-shaped gouge which is used to create channels (or flutes).

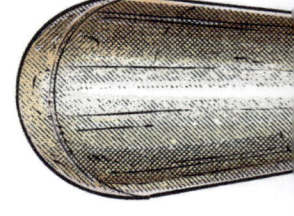

Tool No. **83**

SPOON GOUGE

So called because of its distinctive spoon shape, this gouge comes in a wide range of shapes and sizes, although the shanks are usually quite long. They're really useful for getting deep into a hollow (and other hard-to-reach areas such as inside tight curves) with less force than a regular gouge and scooping the wood out.

A back-bent spoon gauge is very similar but the "bowl" part of the tool is upside-down. You can also get spoon gauges with angled cutting edges skewed either to the left or right, to get into tight corners.

Both J.B and S.J. Addis (see page 132) produced fine spoon gouges, as did Herring Bros. In 1850, 16-year-old Thomas Herring, the son of an edge toolmaker in Sheffield, moved down to London and started working for Samuel Joseph Addis. He was clearly a popular lad because he ended up marrying Samuel's daughter three years later. Thomas set up shop with his brother Edwin across the road from Samuel J. in Gravel Street, Southwark, south London. They went their separate ways, but it was Thomas's relations who went on to set up Herring Bros.

One gouge-making company that still survives today is Henry Taylor, which was founded in 1834. Like James Addis, Henry Taylor also won a Prize Medal at the Great Exhibition in 1851 and the 1862 Great London Exposition. And the company is still based in Sheffield.

Spoon gouge

As the name suggests, this is the gouge you need for scooping wood out of deep hollows and tight curves.

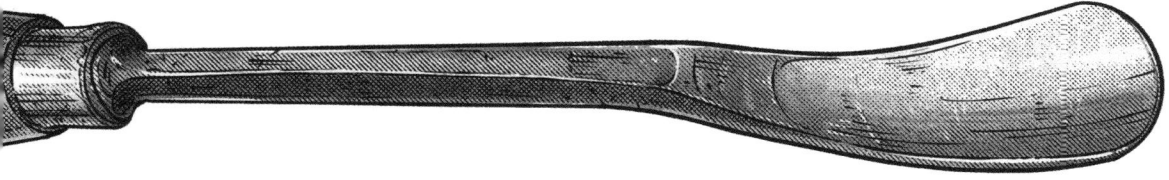

CHAPTER 6
FILING, SMOOTHING & SHAPING

FILES

A file is a toothed tool used mainly to shape or smooth metal and wood, but also, in modern times, plastics and composite materials. Files come in a wide range of shapes, sizes and arrangements of teeth depending on what material you're working with and the finish you're after. They're ancient tools and would have been used in the Stone Age to sharpen other stone tools.

By the Middle Ages, steel files would have been hand-forged by blacksmiths, who cut the teeth individually with a chisel and hammer before hardening the metal by heating it up in a coke-fired oven and quenching it in cold water. This process would have taken some time, which was perhaps one of the reasons why Leonardo da Vinci designed a file-cutting machine in his Atlantic Codex, the huge, 12-volume collection of writings and drawings that he worked on from 1478 to 1519. As with so many of his genius ideas, though, the file-cutting machine was never produced.

It seems that by the 16th century, Germany had become the centre of file production, centring in the towns of Nuremberg, Leipzig, Cologne and Augsburg. Sheffield, south Yorkshire, followed in the 1700s, and by the 1760s, the quality of the files produced there, made from crucible cast steel, had earned them an international reputation. Specialized filecutter's hammers were used, forged from wrought iron with the face made from steel, with a shaft set at around a 60° angle to the head. The hammer head itself was an amazing-looking object with a big hump three-quarters of the way towards the face, which was an almost triangular shape.

Filecutters would use these hammers to strike a small chisel thousands of times on a file "blank" to shape the surface with teeth. This process was done with such amazing speed that filecutters became known as "nicker-peckers", a local term for a woodpecker!

Files were still handmade in the UK until the late 1800s, but elsewhere huge leaps forward were being made. In Switzerland, F.L. Grobet, who had started his toolmaking business in the town of Vallorbe in 1812, invented a file-cutting machine in 1836. It couldn't have come along at a better time, because Swiss watchmakers, who needed files to make and repair their timepieces, were then producing well over a million watches a year. By 1850, that number had reached 2.2 million. The company that F.L. Grobet founded merged with two other family businesses in 1899 to form Usines Métallurgiques de Vallorbe, which is still producing some of the best files in the world.

A file's "cut" tells you how the teeth on the face are arranged. There are quite a few of these, but we'll go into the three main ones. First you've got single-cut, which are parallel rows of teeth that cut in one direction and are usually set at a 65° angle. Next you've got double-cut files, which have two sets of parallel lines that form a diamond pattern. This one cuts quicker than a single-cut file and is used mainly for rough work. A curved cut file has curved parallel rows of teeth that stop the file from clogging up. These ones are often used for car-body repair.

Then you've got the various grades of coarseness of a file, which relate to how the teeth are spaced and the number of teeth per inch on the face. As with sandpaper, though, start with the coarsest and work through to the finest to get the smoothest result. And remember that files cut on the push stroke (the forward stroke). Trying to file on the pull (return) stroke can dull the teeth and damage the tool.

HALF-ROUND FILE

A half-round file is often used on sheet metals and wood to smooth down material you've cut, and for sharpening tools. It has a semi-circular cross-section, so one side of the file is flat and the other half-round.

This means that half-round files are great for working on corners and curved surfaces. The end of the blade often tapers slightly both in terms of thickness and width towards the point, so they're good for getting into small areas, such as the inside of a pipe, although some half-round files keep an even thickness and width to the tip. The teeth on the rounded side can either be single cut or double cut and they're usually double cut on the flat side.

A half-round ring file is narrower than a normal half-round file, and they're usually 150mm (6in) long. They taper in thickness and width to a point and are used for filing the inside of rings.

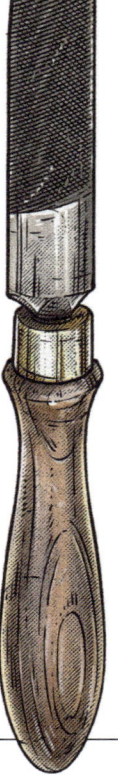

Half-round file
Their half-cylindrical shape makes these files ideal for descaling pipes.

Tool No. **85**

MILL FILE

This common type of file is rectangular in cross-section and can either have even thickness or taper slightly in width. Mill files are single cut and are used for general shaping of metal and wood and for sharpening flat-bladed tools and different types of knives.

You can also use mill files to file saws to keep the teeth nice and even height-wise. If you see the word "bastard" on a mill file and are a little surprised, I wouldn't blame you, but it's a measure of the file's coarseness. "Bastard" is a middle-grade of coarseness that sits between "coarse" and "second cut", so it has quite coarse teeth that are good for removing material quickly. So, how on earth did it end up with that name?! I've come across all sorts of theories, but the one that seems to ring true (to me at least) is that it's related to the "bastard sword" – a term that cropped up in the Middle Ages for a sword that, size-wise, sat between a long sword (a really big sword that you'd use with two hands) and an arming sword (a shorter, lighter one-handed sword). A bastard-cut file sits in the middle of the coarseness stakes in a similar way.

The similar flat file can be single cut or double cut, with double cut being better for the initial heavy-duty shaping process.

Mill file

Mill files have most commonly been used over the years to sharpen other tools, and are so called as they would have been used to sharpen mill saw blades.

RASP

A rasp is a perfect tool for the preliminary work of shaping and sculpting wood. It looks very much like a file but has coarser teeth, which are made by individually punching the teeth into the tool either with a machine or by hand, a process known as "stitching". Although, having said that, there are only a few rasp-makers who make their tools by hand anymore. One of the most famous is Michel Auriou, a fourth-generation rasp-maker based in Saint-Juéry in the southwest of France. The history of the business dates back to 1856 in Paris before Michel's grandfather moved to Saint-Juéry in 1933, a town that once contained hundreds of rasp-makers.

Michel has a room where his blacksmiths forge tool blanks and grind down the cutting edges before they move on to the "stitching room", where the rasps are given their teeth using a "barleycorn" pick and hammer over an anvil. All in all, there are 30 separate operations going on to make each rasp, and for the finer-cut rasps, they can take 90 minutes each to painstakingly stitch (and that's by a master stitcher). But Michel's commitment to his craft gives a major advantage over machine-cut teeth, and that's because the machine gives you perfectly uniform rows, which isn't really what you want, because you end up with grooves in your workpiece. Hand-stitching the teeth creates a random pattern, which gives you a much smoother finish. And that's something that speaks to my heart, and is one of the reasons why people like me get so passionate about hand tools – unique techniques passed down from generation to generation that are still creating truly unique tools.

Tool No. **87**

RIFFLER

It sounds more like a jazz guitar than a hand tool, but a riffler is a small, slim, double-ended file. I actually only found out that this specialized file had a specific name when I started working on this book. The unusual word actually comes from the Old French word *rifler*, which means to scrape, and that makes sense because they're used for shaping and smoothing metal, stone and wood, especially in difficult-to-reach areas. Rifflers have cutting surfaces on their two heads, while the middle section is uncut and is used as a handle. They're produced in lots of different sizes, shapes and cross-sections, and the ends of rifflers can look quite different, with curved, straight and knife-like cutting surfaces.

There are two different sorts of rifflers: silversmith's rifflers and die-sinker's rifflers. Die-sinker's rifflers are smaller and finer than silversmith's rifflers, which tend to have longer handles and wider ends. In case you're wondering, a die-sinker is someone who engraves dies to stamp coins, medals and similar objects with. Die-sinker's rifflers are also often used by jewellery makers, musical instrument makers, wood carvers, sculptors working on delicate features and even vintage car and motorbike restorers.

I've got about 20 rifflers in my barn at *The Repair Shop* that I bought at the Beaulieu International Autojumble a few years ago. I remember seeing them on a table at a stand with an elastic band bundling all of them together and thinking they'll be handy. I don't use them often, but when I do, they're so useful because nothing else can really do the job, getting into narrow recesses and curves.

The best in the business are Auriou (see opposite), who started producing octagonal middle sections rather than the standard square to make them more comfortable back in 2007; and Vallorbe (see page 143).

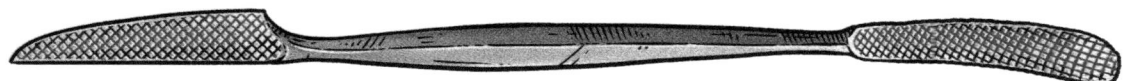

Tool No. **88**

NEEDLE FILE

Needle files are a category of thin files suitable for precise, detailed work, which is why they tend to be used a lot in jewellery making, watchmaking, engraving and by model makers all over the world. You can get them in all sorts of different cuts and shapes, including half-round, square, round, triangular and the more obscure-sounding ones, such as warding files. These were originally designed to work on wards – obstructions in a lock that stop the lock from opening unless you've got the right key.

The needle file is one of the 22 files listed in Smith's Key from 1816, one of the earliest illustrated catalogues of tools (and cutlery) being produced in Sheffield at the time. But we've talked a lot about Sheffield in this book (for good reason, though!), so it's time we spent a bit of time over the Yorkshire county border and into Lancashire.

One of the filemakers who would have been working at the time Smith's Key was being drawn up was Peter Stubs, who started up a small filemaking business in Warrington, Lancashire, in 1777. Eleven years later, he took over the Warrington pub that he was born in, the White Bear, and seems to have managed both

businesses. The filemaking side of things must have been going well, though, because he approached Joseph Smith (who would later go on to become famous for producing Smith's Key) to engrave and print an illustrated 55-page catalogue of their tools in 1801.

The following year, Stubs sold the pub and established a filemaking factory nearby, a brave move at the time, but Stubs clearly knew what he was doing and was soon employing over 40 people.

They concentrated on producing files for watchmakers and repairers, and needle files would have been go-to tools for anyone in those professions.

The needle file is one of 22 files listed in Smith's Key from 1816, one of the earliest illustrated catalogues of tools.

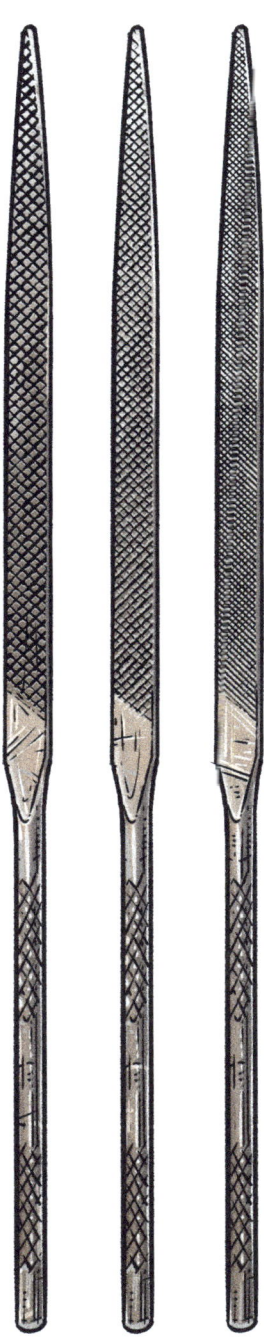

Needle files

Needle files were once made entirely of steel (hence the Sheffield connection) but are now available with ergonomic soft-grip handles, making them easier on the hands.

Peter Stubs died four later, but his two sons carried on the family business, and by the middle of the 1820s, they were producing their own steel. They later forged an international reputation for their silver steel, the round metal bar steel supplied to toolmakers. Confusingly, it doesn't actually have any silver in it, but it looked like silver because it has a polished finish. Although the company passed out of the Stubs family's hands in the 1840s, the firm continued to use the very well respected Peter Stubs name.

One common problem with vintage needle files is that their very slim knurled handles can cause finger cramping if you're using them for a long time. Thankfully, specialized file holders have since been produced to spare people's hands.

PLANES

Used for shaving down, shaping and smoothing wood with an angled blade that you push across the wood's surface, planes probably date back to ancient Egypt. Plane-like tools appear in hieroglyphics and wooden items found in Egyptian tombs seem to indicate that their surfaces had been shaved and smoothed by a plane-like object.

The earliest-known planes to have been recovered are Roman, found at Pompeii in the 19th century. Almost all of the Roman planes that have been found are smoothing planes, made from iron and wood, except for one example found in Yorkshire in 2000, with an ivory body, and one in Cologne, made from bronze. The blade of one Roman plane, which was found at the bottom of a well in a Roman fort in central Germany, was set at a 50° angle, not far off the sort of angle you find in a modern plane. In terms of function, they're not far off modern planes at all, although the wooden wedge that kept the blade in place was commonly replaced in the middle of the 19th century with the screw and lever adjustor.

There are many different types of plane, but the three you'd commonly use if you're starting on a sawn board by hand is a jack plane, a jointer plane and a smoothing plane. These three will take you from roughing out a sawn board to a finished surface. There's also the scrub plane (used for heavy-duty wood removal), fore plane (a bit longer than a jack plane and used for larger workpieces) and block plane (used to cut at right angles to the grain and for finishing work), as well as at least 20 more specialized planes of all different sizes.

One company that's had a big impact on planes is Stanley, so much so that its numbering system is still used by other manufacturers. Smoothing planes are numbered 1 to 4½, jack planes are 5, 5¼ and 5½, the fore plane 6, and jointer planes 7 and 8. The system follows a rough size order – the higher the number, the bigger the plane.

Tool No. **89**

SMOOTHING PLANE

After you've finished with the preparation work, this plane will take you to a nice, smooth finish. And if you can master this tool, you won't even need to use sandpaper. It's also useful for trimming joints, cleaning up edges and getting rid of little marks left by other tools.

Smoothing planes range from 140mm (5½in) to 250mm (10in) in length but the most common is the No. 4, which is 230mm (9in) long. The No. 1 was produced by Stanley from 1869 to 1943, and while very small and not often used much these days, they are beautiful tools that tend to go for a lot of money when they come up on auction sites.

Old Stanley planes (and other Stanley tools, for that matter) have the "sweetheart" trademark – a heart-shaped stamp with the initials S.W. inside. Tools with this stamp were produced between 1920 and 1935 in memory of William Hart, the former president of The Stanley Works. Hart had joined Stanley in 1854, aged 19, and became a member of the board of directors just two years later. He was the president from 1884 to 1915, during which time he oversaw all sorts of pioneering strategies that helped turn Stanley into a powerhouse. Stanley reintroduced the Sweetheart trademark on its planes and other tools in 2008 for a time.

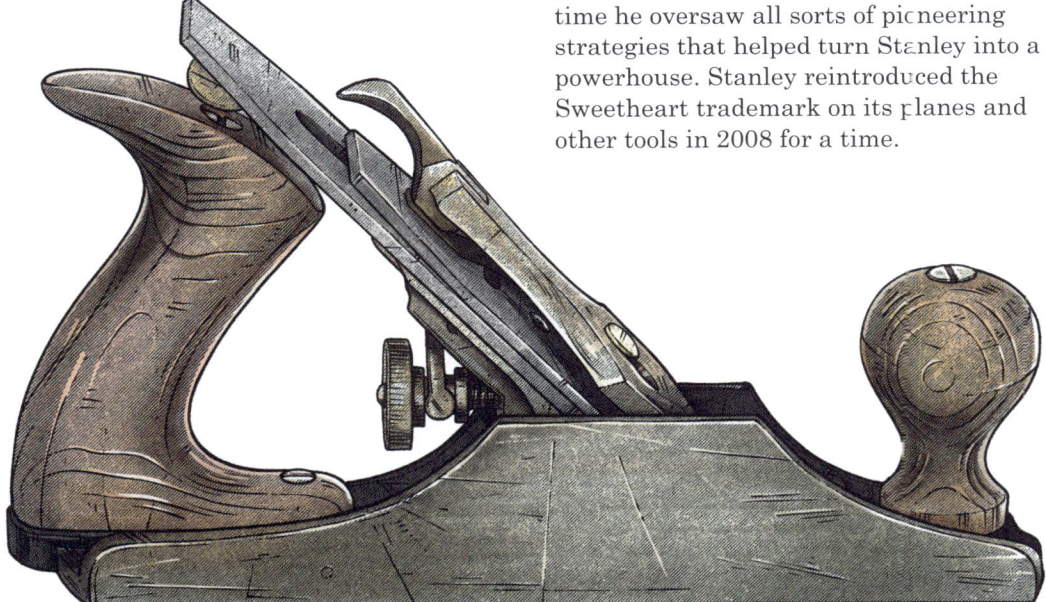

Tool No. **90**

JACK PLANE

The jack plane is a versatile general-purpose plane, with the blade usually set at a 45° angle and a metal body usually made from cast iron (but ductile iron is also used by one toolmaker, who we'll come to). The depth of the blade is adjusted with a thumbscrew that you rotate. The jack plane is the iconic tool that pops into your head when someone mentions the word "plane" – it's the one you'd expect to see in a kids' toy version of a DIY set.

Until the 19th century, the body of a jack plane would have been made of beech wood (in Europe and North America), before iron bodies took over. Leonard Bailey, who was born in New Hampshire in 1825, was one of the toolmakers at the cutting edge of plane-making during the 1850s to 1880s, and his patented 1867 plane design was the forerunner of the modern plane. It's him we have to thank for the adjustable "frog" – the iron wedge that holds the blade at the right angle.

Bailey was a cabinetmaker before he went into the tool-making business in the 1850s. He sold his company – Bailey, Chaney & Co – to Stanley Rule and Level Co in 1869, so Stanley held the exclusive rights to Bailey's designs and the use of the Bailey name. The relationship lasted for only six years, though. Stanley went on to keep producing planes based on Bailey's 1867 patent, while Bailey established a new company to produce another line of planes under the trademark "Victor". Stanley bought this company in 1884 but stopped production of the newly acquired

brand not long after. Bailey ended up retiring from plane-making altogether, but his name sill appears on a number of Stanley's planes today, including its Professional Jack Bench Plane.

So who's Jack, then? It's a bit of guesswork, but "Jakke" was used as a general term by the Normans to refer to the common man, and, over time, the name "Jack" became linked to a number of different common occupations and objects, such as jack tar (meaning sailor), jack of all trades and jack knife, to name a few.

"Jakke" was used as a general term by the Normans to refer to the common man and, over time, the name "Jack" became linked to a number of different common occupations and objects.

Jack plane

This is the tool we tend to think of when someone says, "pass me the plane". It has many uses and its iconic design has changed little over the years.

MOULDING PLANE

This specialized type of plane is used to shape wood for mouldings. A moulding (also called coving in the UK as well as in Australia and New Zealand) is a curved or shaped feature that covers the join between a wall and a ceiling. Wooden mouldings are everywhere, and the chances are, if you look around the room you're in, you'll see at least one type, most probably a skirting board. But there are dozens of different types, with both the body and blade of the tool shaped as per the moulding they were designed to make.

Two of the earliest-known British commercial plane-makers were Thomas Granford and Francis Purdew, both working in London towards the end of the 17th century. Before then, a blacksmith would have made the blade for a plane, and the wooden parts would have been crafted mainly by the carpenters who used them. Some complicated jobs would have required dozens of moulding planes to be made, but not

many from the beginnings of commercial plane-making remain, so when they come up at auction, they fetch a high price.

Someone was clearing out their grandad's shed recently and found an old wooden box they thought I'd like. And they were right! Inside was a beautiful set of moulding planes with lots of different blade profiles. It's almost like the Swiss Army knife of planes, and every once in a while, it's going to be a godsend.

JOINTER PLANE

So called because it's used to flatten wood in preparation for joining components together, the jointer plane is a big plane – up to 610mm (24in) long, although some wooden-bodied examples are larger – and is used mainly on larger, longer boards. Although, having said that, Alan Peters, the legendary British furniture maker who died in 2009, was famously known for using his No. 7 plane for almost everything he made.

Unlike a jack plane (see page 152), which often has rounded corners on its cutting edge, a jointer plane's blade is straight, so it can cut flat edges. One toolmaker who is making some of the best jointer's planes around is also one of the newest. Lie-Nielsen only started up in 1981 in Maine,

USA, but they've made it their mission to revive discontinued but useful tools to inspire woodworkers and other artisans. And I really appreciate what they're doing there, having tried to do a similar thing myself (albeit on a smaller scale) on my Ranalah project (see page 5).

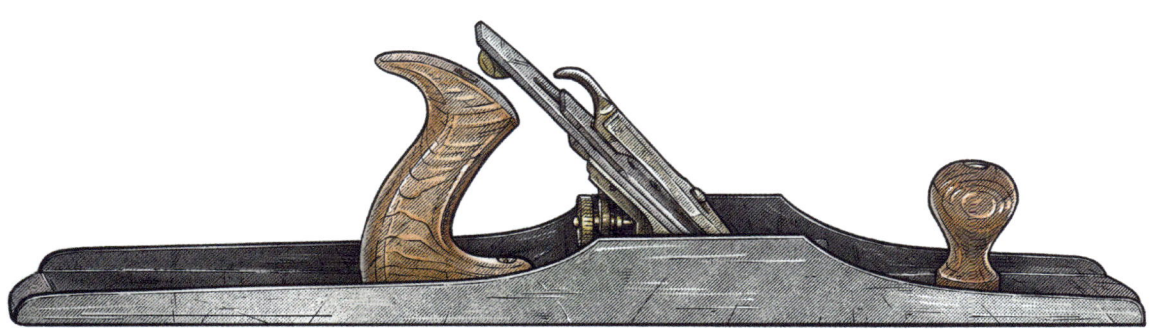

JAPANESE PLANE

Beautifully simple by design, Japanese hand planes (known as *kanna*) are made from white or red oak and most take the shape of a rectangular block. The blade is kept in place with a holding pin, and to change the cutting depth of the blade, you just tap the end of it with a mallet.

There are many different types of Japanese hand plane – ones for roughing out, equivalents to the jack plane (see page 152), planes for smoothing and many more specialized planes besides. They all cut on the pull stroke (i.e. bringing the plane back towards you) rather than on the push stroke, which takes some getting used to for Western craftspeople.

One of the most renowned Japanese plane-makers was Chiyozuru Korehide (1874–1957). Chiyozuru was born with the name Hiroshi Kato and came from a proud lineage of swordsmiths who served the Uesugi family, one of the most famous warrior clans in Japanese history. He became a blacksmith aged just 11, served as an apprentice to his bladesmith uncle

and took the name "Korehide" in his honour after his uncle's death. After establishing a reputation for the quality of his tools, Kato adopted the signature "Chiyozuru" as a nod to the legend of a crane flying over Chiyoda Castle (now part of the Imperial Palace in Tokyo).

The story continues with a young bladesmith by the name of Kanki Yoshiryo (1908–99), who was born in Miki City (16km/10 miles north of Kobe), which is well-known for its tradition of tool making. One day, while he was in Osaka, he came across an exquisite *kanna* (plane) that had been made by Chiyozuru Korehide. So overwhelmed was he by the quality of the craftsmanship that he begged Chiyozuru to take him on as an apprentice. Chiyozuru eventually agreed, and Kanki learned his master's craft. Twenty years later, he succeeded his master, earning the signature Chiyozuru Sadahide from him.

Sadahide paid back his master's faith-winning awards for his blades and a Medal of Honour from the government in 1982. Sadahide created a famous plane which he called *Awaji no Yunagi* ("Evening Calm on Awaji Island"), that puts some of the West's boring tool names in perspective! Made with a sword-quality blue steel blade, this tool is widely regarded as a masterpiece.

Sadahide's son (born Kanki Iwao in 1944) learned the craft from his father and took over the mantle in 1990, but his succession needed to be approved and recommended by many people, including toolmaking masters. Like his father, Sadahide II won all sorts of awards and honours, and trained an apprentice, who received the name Chiyozuru Sadahide (III) in 2019. The craft of creating hand tools with respect for time-honoured tradition is still very much alive in Japan.

The craft of creating hand tools with respect for time-honoured tradition is still very much alive in Japan.

DOLLIES

A dolly is a solid piece of metal with one or more curved faces. They're used to shape other bits of metal, but are probably most often used when repairing car panels. You hold the dolly on the opposite side of the metal you're hammering, to shape the metal to match the curve of the dolly. You can also use dollies on their own as a small hammer in areas that would be difficult to swing a hammer about, such as behind the bumper of a car.

The shape and curvature of dollies varies massively, but there are several that are commonly used: a toe dolly, a heel dolly, a wedge (or comma) dolly, a general-purpose dolly and an egg dolly (see opposite and on page 160). You tend to find the first four of these when you buy a body hammer and dolly tool set. Dollies are normally made from carbon steel, which is then heat treated and sand blasted before being ground and polished, so you end up with a surface that you should be able to see your face in clearly.

I've got a collection of dollies that I picked up from a retired metalworker who I met many years ago when I was looking to buy my first English wheel (also called a wheeling machine), a machine used to curve metal panels. Wheeling machines were first made in the 1890s but only became known as English wheels by American industrialists during the Second World War, when they were used to make aircraft panels. In the 1930s, the coachwork company Ranalah started making them and they're widely considered the Rolls Royce of wheeling machines.

The retired metalworker wanted £3.5k for the English wheel, which was cheap, but right then, I just couldn't afford it. So I just stood there gazing longingly at it. And at the other tools and equipment he'd built up over his 50-plus years career. He'd started as an apprentice for a coachbuilder in west London before going on to build panels for Routemaster buses. After that

Over time, tools become your friends – you spend so much time with them, they help you out and you come to rely on them.

he built cars, and eventually started his own company making expensive cars, and had several guys working for him. When he retired, he moved down to Kent, where he had a workshop, but he stopped using at as much as he used to.

The first thing I saw in his workshop was this stunning collection of planishing hammers (see page 67) and dollies. It was like I'd walked into Aladdin's Cave for metalworkers. So I asked him how much he wanted for some of them. He said he'd only let all of them go together, but when I tried to ask him how much he'd charge, he changed the subject. I got the picture that he wasn't ready to part with them, so I didn't push it, but I thought of them often and wondered where they'd end up.

Then, every once in a while, he started listing some of his tools for sale online, but only when he was ready to let something go that he was still very attached to. I'd message him as soon as I could and he'd ask me to meet him at the workshop. In the end, I went there about five times, picking up little bits and bobs each time. And every time, I'd subtly remind him to keep me in mind if he ever thought about selling his dollies and hammers. He said he'd let me know, but to be honest, I wasn't sure I'd ever end up with them. I knew that lots of other metalworking enthusiasts went to his workshop to buy this and that off him, and I knew they would have had the same reaction as me when they saw the dollies and hammers. Eventually, I assumed that someone with more money than me went and took the lot off him.

Then one day, out of the blue, he messaged me and asked if I wanted the dollies and hammers. I went round straight away. When I got there, I walked into a completely empty workshop except for his prized collection of hammers and dollies in a corner. He'd kept to his word. He told me that lots of people had asked if they could have them over the years, but that he'd remembered that I'd asked first, so he'd kept them for me.

As I was loading them onto the van, he got quite emotional, because not only were they tools he'd used all his working life, but also, some of them had belonged to his dad. Every hammer he handed over, he'd tell a fond story about it and explain exactly which jobs he'd done with it, before saying goodbye to each one. He kept a couple of them that were too hard to part with, to remind him of his dad and his working life. I completely understood, and think I'll

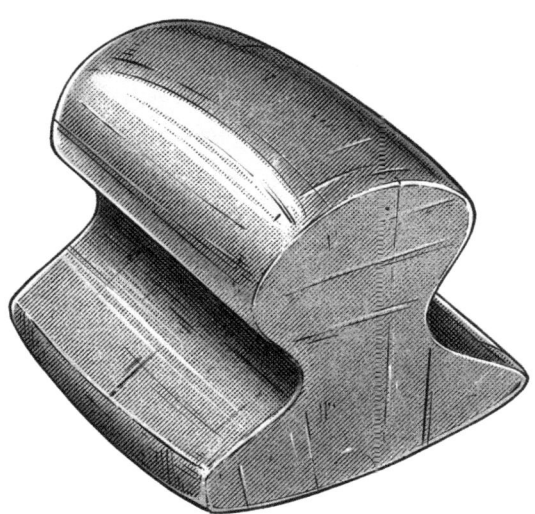

General-purpose dolly
Easy to hold and manoeuvre, the general-purpose dolly has multiple crowned edges, so it can be used for a wide range of auto-repair jobs.

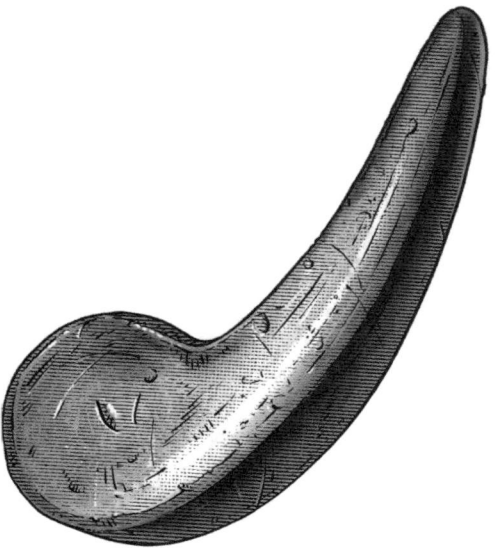

Wedge or comma dolly

This dolly is really helpful for working in tight spaces such as behind brackets on bumpers. The thin edge is good for corners.

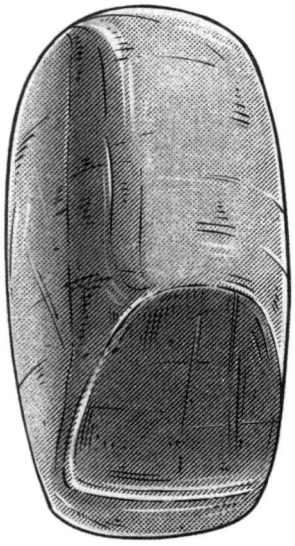

Egg dolly

A versatile dolly with lots of different angles, edges and curves. It tends to get overlooked in auto-repair kits, but it's really useful.

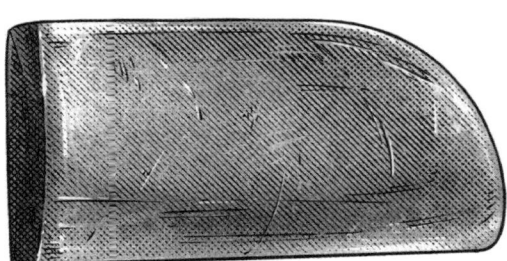

Toe dolly

So called because it looks like the toe of a shoe, you'd typically use it to reshape flat, low-crowned car panels.

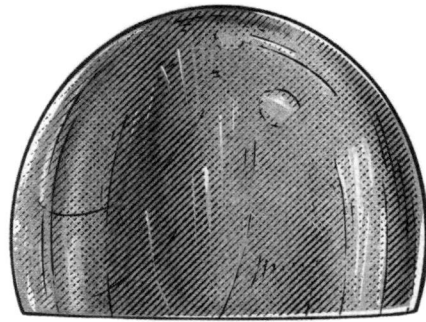

Heel dolly

One side is C-shaped and the other is a flat edge, just like the heel of a shoe. A heel dolly is really useful to help you reach into sharp corners.

probably do something similar when I eventually hang up my welder. Over time, tools become your friends – you spend so much time with them, they help you out and you come to rely on them.

I feel very lucky to have his set of dollies and hammers. It almost seems like it was meant to be that I've become the custodians of them now, and I'm lovingly spending time sanding them down and polishing them up so they'll look great. I think he was only prepared to let them go to someone who would treasure them and understand how important they were to him. He knew I was working on the Ranalah project (see page 5) and understood how passionate I am about it. But it's funny – whenever you meet one of the old boys with a lifetime of experience in building cars using an English wheel, they always say the same thing to you: "It takes 20 years of practice to know how to do it properly, son." By that time, I'll be nearing my own retirement! It took time, persistence and patience to build up that trust with him. I got a message from him recently asking if I'd like an old box of photos of the tools and the cars he'd made with them. He's become a sort of honorary grandfather figure now to me. Going forward, I'm going to be using some of his unique tools, especially the low-crowned dollies, on the bonnet of the Porsche 356 that I'm restoring. It's an honour to give his tools a new life.

You hold the dolly on the opposite side of the metal you're hammering, to shape the metal to match the curve of the dolly.

Slapper

Slappers, or spoons as they're sometimes known, got their name from the slapping motion used to shape metal using a dolly (see pages 158–161) on the other side.

A tool with a name you don't tend to forget, a slapper is mainly used in sheet metalwork.

Tool No. **99**

SLAPPER

A tool with a name you don't tend to forget, a slapper is mainly used in sheet metalwork, especially when repairing the relatively soft metal panels of vintage cars. Also commonly known as a spoon, it's basically a fairly long section of metal that doesn't look far away from a palette knife, shape-wise. They earned their name from the slapping motion you make when using one. You usually use a slapper together with a dolly (see pages 158–161) on the other side of the panel you're working on, to shape the metal to the contour of the dolly. Some metalworkers make their own slappers, particularly if they need something specific, such as a particular curve. They make them out of scrap metal such as old leaf springs (used for suspension in cars).

Slappers have got longer, wider faces than hammers, so they work on a bigger area. Plus, their shape and lack of hard edges means that they're great for smoothing panels without leaving marks. Although most slappers have smooth faces, you can also get rough-faced slappers, which are useful for showing the areas you've covered. I've got them in all sorts of shapes and sizes, including some with faces that are almost bent back on themselves, which are great for getting into recesses. Most slappers are made from steel, but you can also get wooden slappers, which are typically covered with leather. Leadworkers use wooden slappers for roofing. They're chunkier than the metal ones used by panel beaters, but they've got the same flat bottom and they're ideal for patting in the lead that lines roofs.

The company that I've bought almost all of my spoons, or slappers, from is Sykes-Pickavant. It started as J.W. Pickavant & Co Ltd in 1921, a company supplying motor accessories, spares and tools, that was founded by Jack Pickavant. Jack hired a chap called Joe Sykes as the company's first salesman, but he left to start his own tool distribution company in 1931, which he called Sykes. The two companies were eventually reunited, merging together in 1968 to become Sykes-Pickavant. It's a great old British brand that specializes in panel-beating equipment such as planishing hammers (see page 67), dollies and slappers.

CHAPTER 7
TIGHTENING & LOOSENING

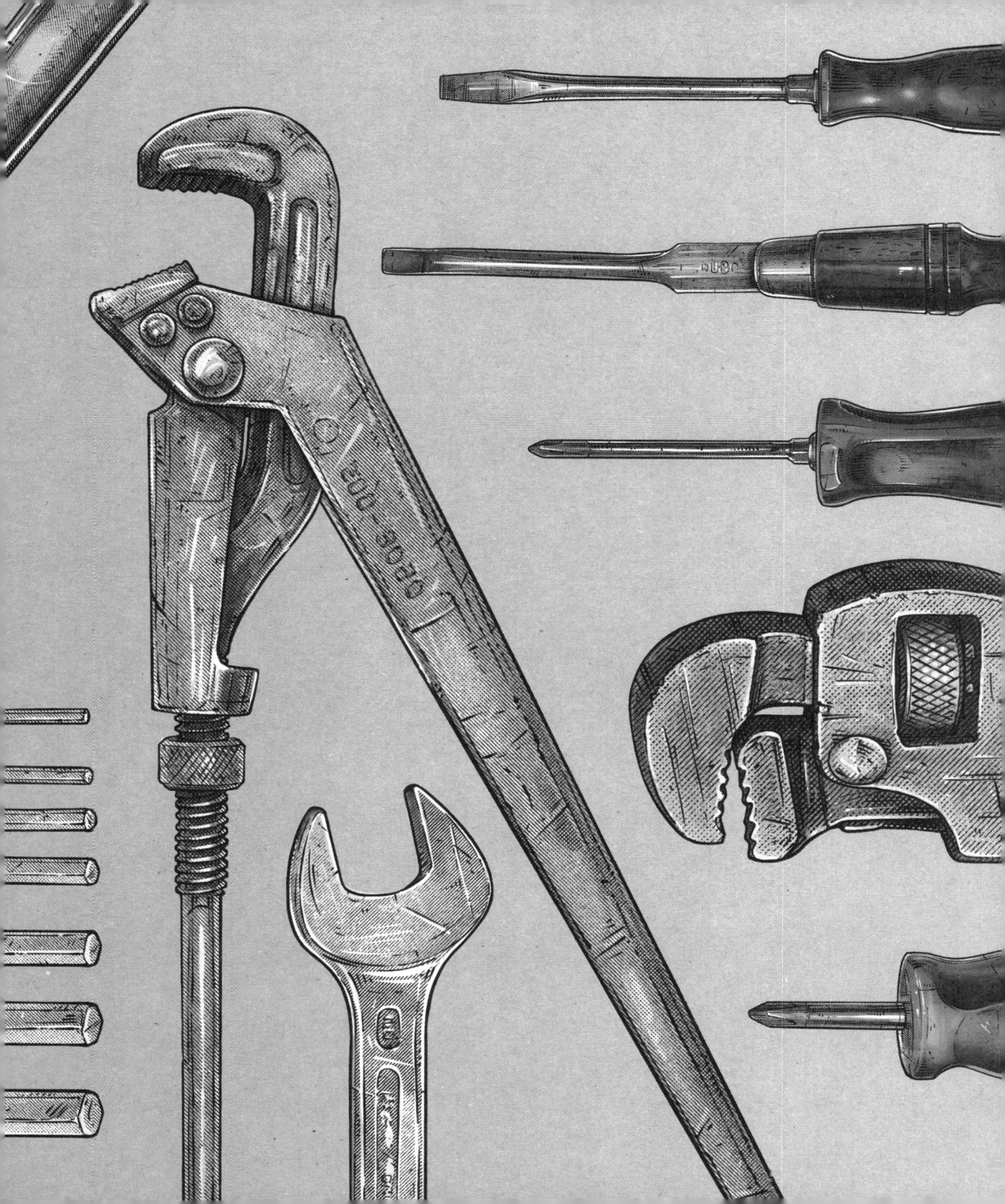

SCREWDRIVERS

Archytas of Tarentum, a Greek philosopher and mathematician, is usually credited with the invention of the screw thread in the 5th century BCE. But exactly when a screw was used in the way that we'd recognize today isn't clear. What we do know for sure is that by the 1st century CE, wooden screw presses were being used to squeeze grapes to make wine and olives for oil. And we know that Romans were using screw presses to press clothes, because remains have been found at Pompeii.

It wasn't until the 15th century that metal screws were used, and this was when Leonardo da Vinci spent some time drawing up a design for a machine that could cut screws. The aim was most likely to cut down on the huge amount of time it took to cut a screw thread by hand. It makes sense that screwdrivers would have first appeared around then, probably hand made by blacksmiths, but we've got little evidence to go on. When they first came about, though, they wouldn't have been known as screwdrivers. In England, it seems they were first known as "turnscrews", which seems very much like a direct translation of the French word *tournevis*.

It wasn't until the time of the Industrial Revolution that screws started being manufactured in factories using early versions of screw-cutting machines, like the one developed by brothers Job and William Wyatt in Staffordshire in 1760. Over the next 50 years, more and more producers started making screws for the rapidly increasing numbers of applications.

And they'd need specialized screwdrivers to tighten them. In the earliest existing price list of Sheffield tools, dating from 1828, there are some fascinating-sounding screwdrivers, including the Gent's Fancy Turnscrew and Sewing Machine Turnscrew. They were available in two styles: Scotch pattern (with a flat blade that tapers gradually from the handle down to the tip) and London pattern (with a slim waist opening out onto a wider flat blade that tapers to the tip). If you look up vintage screwdrivers, you'll see 19th century tools for almost every trade you can imagine: coffinmaker's, cabinetmaker's, gunsmith's, cabinetmakers and clockmaker's, to name just a few.

But as the number of different screws increased massively, one thing became clear: someone was going to have to come up with some sort of standardization plan. Up stepped a man with that plan in around 1841, Joseph Whitworth developed the first national screw thread standard, which became known as BSW (British Standard Whitworth). And these screw sizes are still

Flat head (right and centre right)

Flat-headed screwdrivers have remained relatively unchanged since the early 1800s.

Phillips (left and centre left)

Henry F. Phillips was granted the patent for the Phillips cross-head screwdriver in 1933.

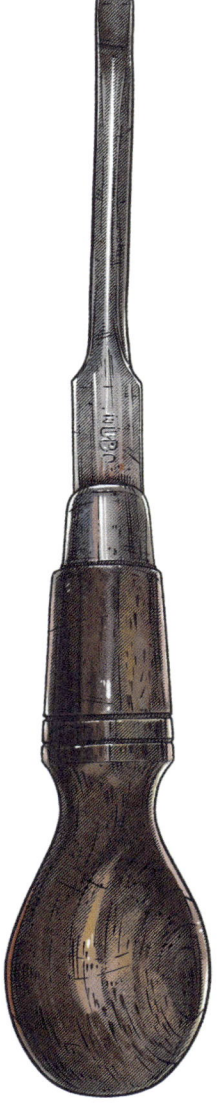

used today by folk like me, who spend their days repairing prized old possessions.

That was until the next major turning point (I had to get one pun in there) happened in 1907–08 when the Canadian inventor P.L. Robertson (1879–1951) patented the square-socketed screw. It had a screwdriver to go with it, with a square-shaped protrusion. This was a major advance over a slotted screw because it was easier to insert and didn't slip.

Then, in 1932, again in North America, a chap called John P. Thompson applied for patents for a new type of cross-headed screw and the matching screwdriver. Thompson was something of a jack of all trades, but it seems he was working as an auto mechanic in Portland, Oregon, when he came up with the design. I say "new type of cross-headed screw" because the first cross-headed screw to be patented was designed by a British engineer called John Frearson in 1873. Unlike Frearson's first cross-headed effort, though, Thompson's screw had a curve in the centre, which means that once you've tightened the screw as far as it can go, the screwdriver pops out. Clever.

So where does Mr Phillips fit in? Well, Henry F. Phillips was the managing director of a mining company, who entered into a business arrangement with Thompson around 1933. It's Phillips who the patents were actually awarded to. Phillips soon formed the Phillips Screw Company and set about trying to find a manufacturer to produce the screws. He succeeded, and the American Screw Company started making them (after a few more design modifications were sorted and patents were issued to Phillips). In 1936,

Phillips secured his first big industrial customer, GM Motors, which used the new cross-headed screws on its 1936 Cadillac. By 1940, almost every US car manufacturer was using Phillips screwdrivers. And now, pretty much everyone has one in their toolbox. The chances are if you have a look at what you're sitting on, you'll find a Phillips screw somewhere.

The Pozi drive (more commonly known as Posidriv) was an adapted version of the cross-head screw that the American Screw Company and Phillips Screw Company worked on together in the late 1950s. The main aim was to increase the amount of contact areas between the screw and the screwdriver, which would also make the screw less likely to slip. The most obvious visual difference from the Phillips are that the extra grooves that appear in between each of the four parts of the cross and the blunt tip of the screwdriver. There are six different Pozidriv sizes (0,1,2,3,4,5) and you usually see the abbreviation "PZ" before the number. The last thing to mention is that Pozidriv screwdrivers don't work on a Phillips drive, so don't try it!

The last one I want to mention is the Torx screwdriver, which was developed longer ago than you might imagine, back in 1967. With their distinctive star-shaped heads featuring six rounded points, they often get called star bits. There are over 25 different Torx sizes, but the most common ones you'll see are T10, T15 and T25. Torx are better than the Pozidriv when it comes to transmitting torque and reducing slipping (which are major plus points), but they are more expensive and not as common.

Pozidriv (left)

Pozidriv screwdrivers are similar in design to Phillips except that they have a blunt tip and an additional four grooves in between each of the four parts of the cross.

Torx (right)

Torx screwdrivers have a unique six-pointed star-shaped head and are useful for reducing slippage.

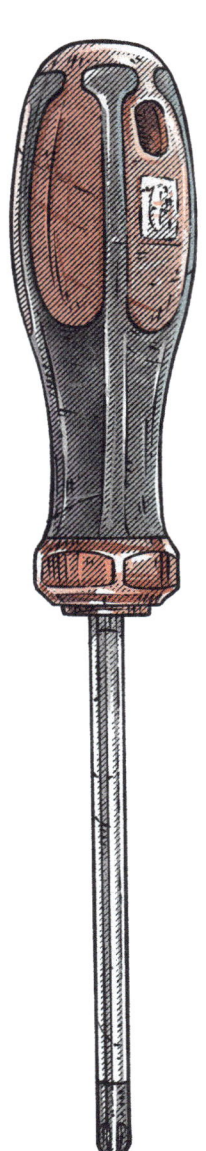
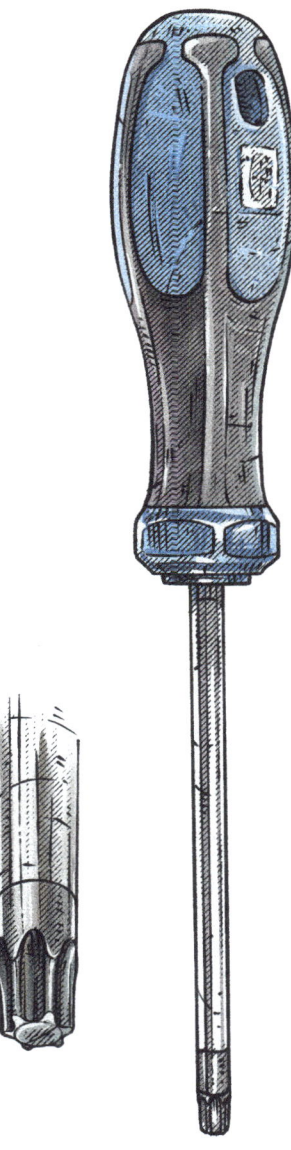

SPANNERS & WRENCHES

So what's the difference? Well it depends where you live. In the UK, Ireland, Australia and New Zealand, they're generally described as two slightly different types of tool – a spanner tightens and loosens nuts and bolts, while a wrench is used to turn things that don't fasten, such as a copper pipe. In North America, they've kept things a bit simpler by generally calling everything a wrench. Well, except for the tool that we in the UK would call a C-spanner; in North America, it's known as a spanner or sometimes a spanner wrench. So we've all got our quirks that confuse each of us, either side of the Atlantic.

As with many early inventions, it was the Chinese that made some of the most important breakthroughs. The Han dynasty spanned from 202 BCE to 220 CE and during that time, they were happily using paper, rudders and even seismographs long before they'd travelled to the West. The historian and author Robert Temple (in his book *The Genius of China* back in 1986) described an adjustable spanner in use in the 1st century with a sliding caliper gauge. Meanwhile, in Europe, we'd have to wait until the 1400s to see an early version of a box spanner, which would have been used to tighten crossbow strings. This development would have been a game-changer war-wise, because suddenly you didn't have to look like Popeye to load a crossbow bolt.

In the 1500s, socket wrenches appeared with T-shaped handles or turning handles that you rotated clockwise to tighten. Both of these types of socket wrenches would have been used for winding up clocks. Meanwhile, box spanners were still being used to hurt people, this time by turning the little wheel on the side of a wheel-lock gun, which wound a chain around a spindle and compressed a spring, which was released by pulling the trigger of the gun. Over the next century, the designs for some of these wheel-lock gun spanners changed from functional to beautiful, decorated with ornate carvings.

The Industrial Revolution changed everything, and things didn't have to be handmade by your friendly local blacksmith anymore. Spanners could be made in factories using cast iron and produced in much larger quantities. During this time, many of the spanners that we're familiar with today would have been manufactured for the first time. Modern high-quality spanners are often made from chrome vanadium steel, which is stronger and tougher than carbon steel.

I've got a soft spot for spanners. When I was young and I had my little six-drawer toolbox, I'd go to car boot sales and pick up random spanners, some imperial, some metric. But when it came to actually needing one of them, it was always a nightmare trying to find the right one. The one that was missing was always the one I needed – it was always the 10mm (⅜in) one!

These days, I've got a strict rule about keeping full sets of spanners rather than odd ones. Looking back, I might have picked up this habit in my early twenties when I worked as the Saturday boy at Karmann Konnection (a supplier of restoration parts and accessories for Volkswagen), watching John the mechanic with his Snap-On toolbox and roll cab, with all his spanners neatly arranged. So now I've got full Snap-On spanner sets (which are great), both in imperial and metric. But I've also had to hunt down spanners that can tighten and loosen the Whitworth and BA (British Association) nuts and bolts, which I find on a lot of the vintage items that arrive on my bench at *The Repair Shop*.

I found a whole bunch of Whitworth spanners in a big bag at a car boot sale, so I laid them all out, made two full sets and gave away the rest of them. Sometimes it's harder to stick to my rule, though, especially when I come across a classic spanner such as a Snail brand one that would have come in a 1950s Rolls Royce toolset!

Tool Nos. 104–105

OPEN-ENDED SPANNER & COMBINATION SPANNER

Say the word "spanner", and you tend to think of a slim tool with U-shaped openings at both ends. This is your classic open-ended spanner, although why the makers of the board game Cluedo decided to turn it into a murder weapon is anyone's guess. In fairness, the spanner from the 1949 launch of the game in the UK does look a lot bigger than you'd expect. In the USA, meanwhile, where the game was also launched in 1949, they came up with the much more menacing-looking monkey wrench (more on that murder weapon on page 178).

Width-wise, standard open-ended spanners range from 3.2mm (⅛in) to 65mm (2½in) to suit the nuts and bolts they'll be tightening and loosening. They each have their jaws set at a 15° angle to the shaft of the tool, so that if you're working in a tight space, you can flip the spanner over and connect with more flat sections of the nut. I've got to say, whoever came up with the 15° angle idea is an absolute genius. Suddenly, gone were the days of turning your spanner a fraction, flipping it over, turning it another fraction and gradually going crazy.

A combination spanner has a U-shaped opening at one end and an enclosed ring or box end at the other. Both ends are usually the same size, and the range of sizes is the same as the open-ended spanner. The ring end usually has six or 12 points that work with hexagonal-shaped nuts and bolts.

You'll recognize the open-ended spanner as a murder weapon from the board game Cluedo.

Combination spanner

The ring end of a combination spanner is designed to work with hexagonal-shaped nuts and bolts.

Open-ended spanner

This classic design has U-shaped openings at either end.

ADJUSTABLE SPANNER

An adjustable spanner is a spanner with a jaw that you can adjust with a screw (or a lever) so it can fit lots of different-sized nuts and bolts. There are a few types of adjustable spanner, including a monkey wrench, which we'll get to on page 178.

The first spanners with an adjustable mechanism were operated by a wedge-shaped adjuster that you hammered into place. It was around 1800 that the first spanner appeared that could be adjusted via screw thread, but a big breakthrough, design-wise, came along in 1808, and we know that thanks to the research of Ron Geesin in his book *The Adjustable Spanner*. William Barlow, who worked at the dockyards in Portsmouth (his employer was the father of Isambard Kingdom Brunel, incidentally) came up with a design for a "wrench for screw nuts of any

size" that featured an adjustable screw and a head angled away from the shaft of the tool. It doesn't seem to have been manufactured, though (except for a specimen), and the world had to wait for another 35 years until someone else stepped up with another design. That someone was Richard Clyburn, an engineer from Yorkshire who in 1843 registered a design for a spanner with the mechanism we're all familiar with: the transverse worm-on-rack mechanism that you operate with a thumbwheel screw to adjust the width of the jaws. This became widely known as "The Clyburn" and was a big international success.

Forty-three years later, Swedish inventor and industrialist Johan Petter Johansson

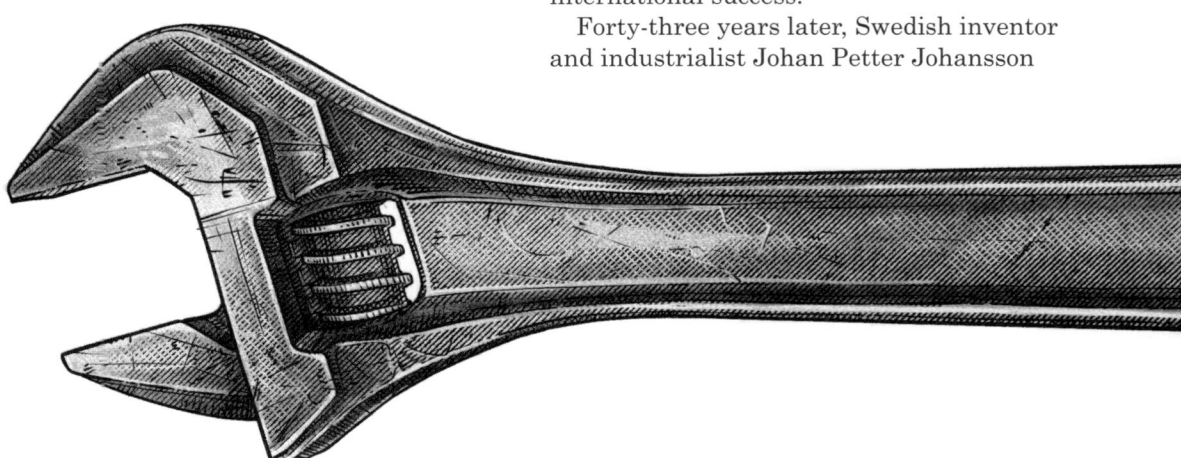

set up his own tool-making company called Enköpings Mekaniska Verkstad (the Mechanical Workshop of Enköping). Four years later, he entered into an arrangement with B.A. Hjorth & Company to sell and market his tools under the trademark Bahco (an acronym formed from the initials of the founder). One of these tools was a modified version of Clyburn's adjustable spanner, which he patented in 1892. It was lighter and more convenient than Clyburn's and became the tool that Bahco is most famous for. In 1914, the company started exporting their adjustable spanners to England, and until 1924, they stamped them with Bahco-Clyburn, presumably a business arrangement to help with marketing.

Johansson's spanner became the inspiration for another design that became so popular in the USA and Canada that the brand name Crescent is now forever associated with the tool. Another Swedish toolmaker was involved, this time a chap called Karl Peterson, who set up the Crescent Tool Company in 1907 in Jamestown, New York. He definitely got something right because the next year, its Crescent adjustable spanner, or wrench, was being supplied along with each new

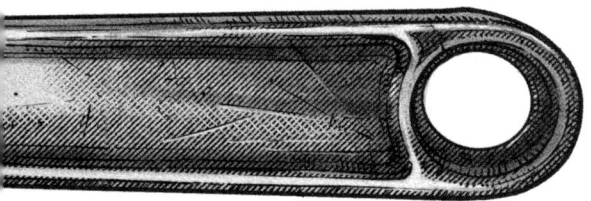

Ford Model T, the first affordable assembly-line-produced car. The Model T car went on to become the biggest-selling car of all time, a record that stood until the 1970s. Nearly 20 years later, Charles Lindbergh gave Crescent a big helping hand by stating that on his famous solo non-stop transatlantic flight in 1927 from New York to Paris, all he took with him was "gasoline, sandwiches, a bottle of water and a Crescent wrench and pliers".

I'm not joking when I say I've got every size of adjustable spanner there is. And I'm not sponsored by them, but they're all made by the Swedish brand Bahco (now part of Snap-On). I had to get a new one on a *Repair Shop* job recently to fix a homemade telescope. The owner worked for the water board and he'd made a telescope out of loads of discarded bits, such as stopcocks, valves, pipes and a massive bolt that must have been from a disused water main outlet. I didn't have a spanner anywhere near big enough to remove this monster bolt, so I had to buy one.

When I had an old campervan, the first thing that went in the back was an adjustable spanner. With classic motors, sometimes it's the only tool that makes sense, like with an old nut that might have got worn, or when someone's filed it down or modified it; an adjustable spanner is the only tool that fits. Yes, there's more chance of rounding off the nut with an adjustable spanner because it'll push on the corners rather than the flat parts of the hexagon, but sometimes you have to do it, especially when you're on the side of the road in your van and you have to change something, pronto! That's happened to me more times than I can remember.

Tool No. 107

MONKEY WRENCH

A monkey wrench is a type of adjustable spanner with wide jaws that open and close by turning a rotating rind set between the lower jaw and the handle. They became very popular in the USA in the 1840s thanks mainly to the efforts of businessman Loring Coes, who developed and patented a new version of the monkey wrench in 1841. The design was inspired by English coach wrenches, used to tighten and loosen the nuts on wagon wheels, which had been imported to North America.

A monkey wrench is one of those tools where the invention story has taken on a life of its own, like the one about the ball-pein hammer being invented by a French mechanic called Jacques Balpein. The monkey wrench tale, which seems to have started in the 1880s, begins in a convincing way, because it tries to bust a myth that a monkey wrench has nothing to do with being "a handy thing to monkey about with". It goes on to tell us that Charles Moncky invented it in the 1850s, sold the patent and used the money to buy a little house in Brooklyn, New York.

While this tale has been widely rejected as a load of rubbish, there does seem to have been a Charles Monk who lived in Brooklyn and made tools. We know that because his age, address and occupation are in the US Census from 1880.

But before we ask Mr Monk for our forgiveness, it seems that "monkey wrench" appeared in print in 1826, when a bricklayer from Chester, UK, called William Darlington was arrested and charged for stealing a "piece of iron called a monkey wrench". By the 1840s, "monkey wrench" is listed in official parliamentary documents as a tool that engine men working on trains should have with them. So the Mr Moncky story isn't true, but whoever made it up had the decency (and possibly sense of humour) to base it on something true to excite (or annoy) people 140 years later.

So why's it called a monkey wrench? This is the part where I confidently tell you that... I don't know. No one does for sure. The best guess I've come across (from a great online article by Peter Jensen Brown) is that it's got something to do with a "monkey on a stick" toy that became very popular in the 19th century. The monkey has its arms and legs wrapped around a stick and shimmied down when moved to the top and let go. Not only does the toy look a bit like a monkey wrench, it also works like one too, because the jaws of the wrench move up and down, so this seems plausible to me.

No one knows for sure where the monkey wrench got its name, but it might have something to do with the "monkey on a stick" toy that became very popular in the 19th century.

Tool Nos. 108–109

PIPE WRENCH & PLUMBING WRENCH

A pipe wrench is basically an adjustable wrench with serrated jaws that grip on to pipes. Like the adjustable spanner (see page 176), this tool is going to take us on a journey to the USA and Sweden.

Daniel Stillson was born in 1826 in Durham, New Hampshire. He worked as a ship's mechanic during the American Civil War, before becoming an engineer for J.J. Walworth & Co, a company based in Cambridge, Massachusetts, that had started out fitting steam heating systems to buildings. Later on it began producing its own pipe fittings and valves. While he was working for them, Stillson came up with an idea for a tool designed to grip on to metal pipe with slanted teeth facing in opposite directions, which made for a stronger grip. He put together a wooden prototype, which his employers were impressed by. They wanted him to prove how strong it was, so they had a version made from steel and asked Stillson to see if it could tear through a pipe, which it did.

Amazingly, they didn't screw him over, but instead insisted that he patent his new tool (which he did in 1869) and license it to the company. They began producing it, and the timing couldn't have been better because heating systems were being installed everywhere and there was a gap in the market for a versatile plumber's gripping tool. The licensing arrangement with J.J. Walworth & Co made Stillson the equivalent of just under $2 million (£1.5 million) in today's money. Not bad. His

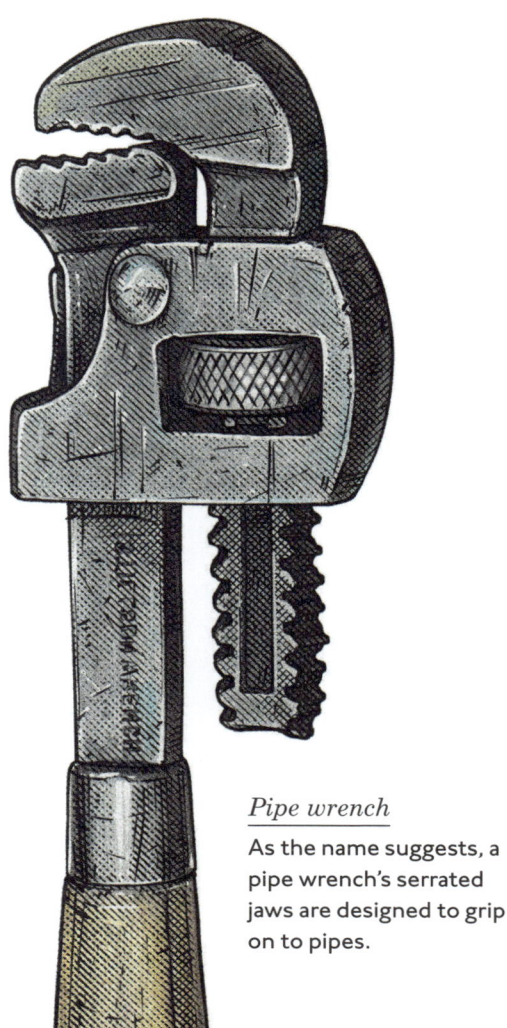

Pipe wrench

As the name suggests, a pipe wrench's serrated jaws are designed to grip on to pipes.

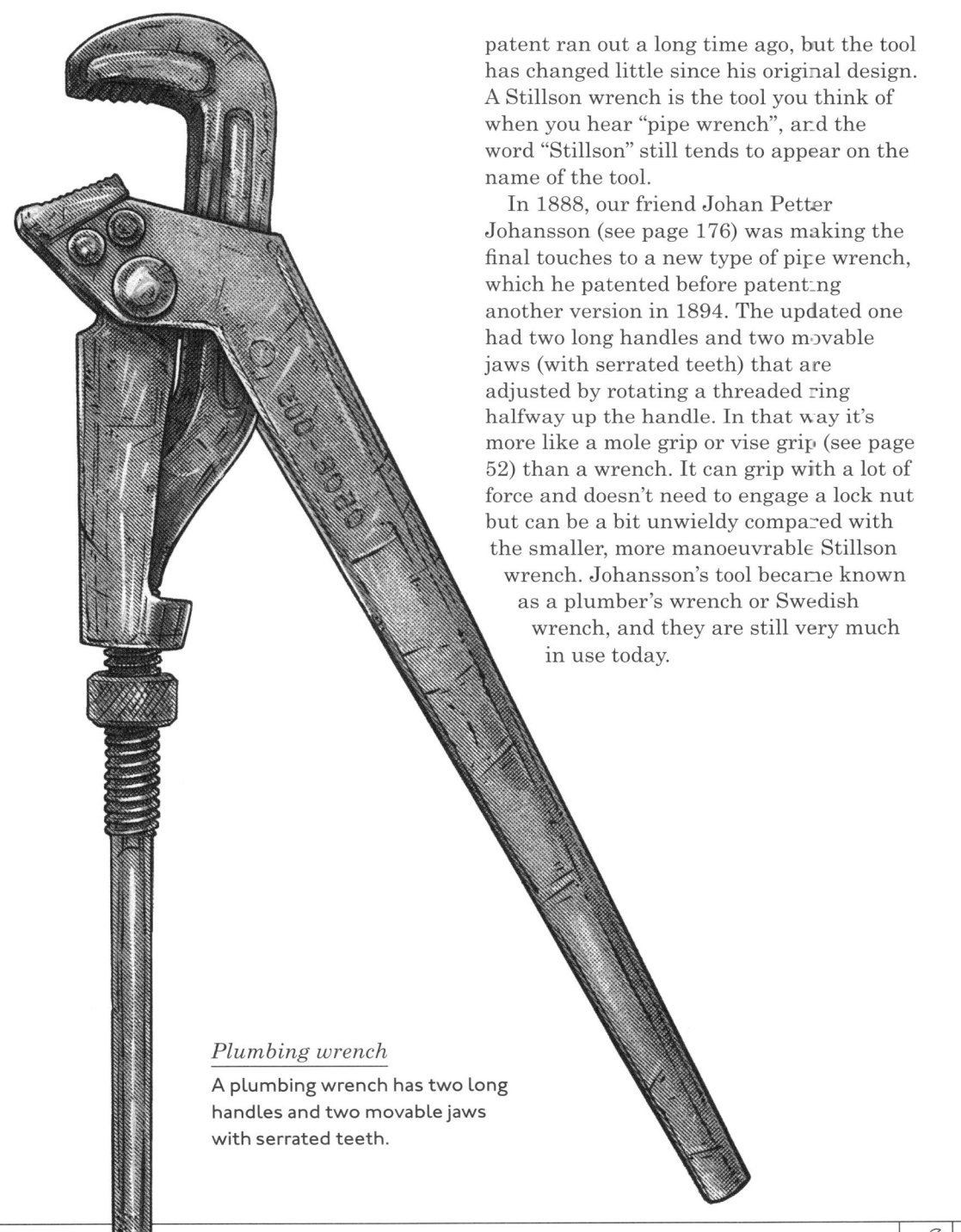

patent ran out a long time ago, but the tool has changed little since his original design. A Stillson wrench is the tool you think of when you hear "pipe wrench", and the word "Stillson" still tends to appear on the name of the tool.

In 1888, our friend Johan Petter Johansson (see page 176) was making the final touches to a new type of pipe wrench, which he patented before patenting another version in 1894. The updated one had two long handles and two movable jaws (with serrated teeth) that are adjusted by rotating a threaded ring halfway up the handle. In that way it's more like a mole grip or vise grip (see page 52) than a wrench. It can grip with a lot of force and doesn't need to engage a lock nut but can be a bit unwieldy compared with the smaller, more manoeuvrable Stillson wrench. Johansson's tool became known as a plumber's wrench or Swedish wrench, and they are still very much in use today.

Plumbing wrench

A plumbing wrench has two long handles and two movable jaws with serrated teeth.

SOCKET WRENCH & TORQUE WRENCH

We've covered the socket wrench (see page 172), but only got as far as the 16th century, so there's a bit of catching up to do. Modern socket wrenches engage with a ratchet, allowing you to tighten or loosen a nut without having to remove the wrench to fit it back onto the nut.

You just need to choose the right socket size for the job, and fit it to the ratchet. This makes the process of tightening and loosening much quicker, allows you to work in a confined space and also means you don't have to carry around a set of non-ratcheting wrenches.

J.J. Richardson, an American inventor, patented the ratcheting socket wrench (usually just called a ratchet) with changeable sockets in 1863. He applied for the patent through *Scientific American*, a magazine that has been going since 1845 and established the first branch of the US Patent Agency in 1850. The magazine encouraged new inventions and even supplied technical help and legal assistance to inventors.

Different-shaped nuts, including the hex (hexagonal-headed) nut, were being mass-produced from the 1880s. The great advantage of the hex head was that you didn't have to turn the socket wrench as much. Then in the 1930s, a New York City Water Department employee called Conrad Bahr came up with a device that enabled you to tighten fasteners with the same amount of torque. In 1935, he (along with his business partner George Pfefferle) patented the adjustable ratcheting torque wrench that gave audible cues and stopped you from ratcheting backwards once the right torque had been reached.

For years I had an old mechanical torque wrench, but I bought a digital version a few years ago. When you're doing something like rebuilding a car engine and every bolt has got to be absolutely right, it's a reliable bit of kit. It's not something I use every day – it's all singing and dancing and a bit pricey – but it is very handy.

Socket wrench (left)
Traditional socket wrench sets come with a range of sockets for working with different nut sizes.

Torque wrench (below)
Dial torque wrenches indicate the torque (the force needed to rotate an object) required to loosen or tighten a nut.

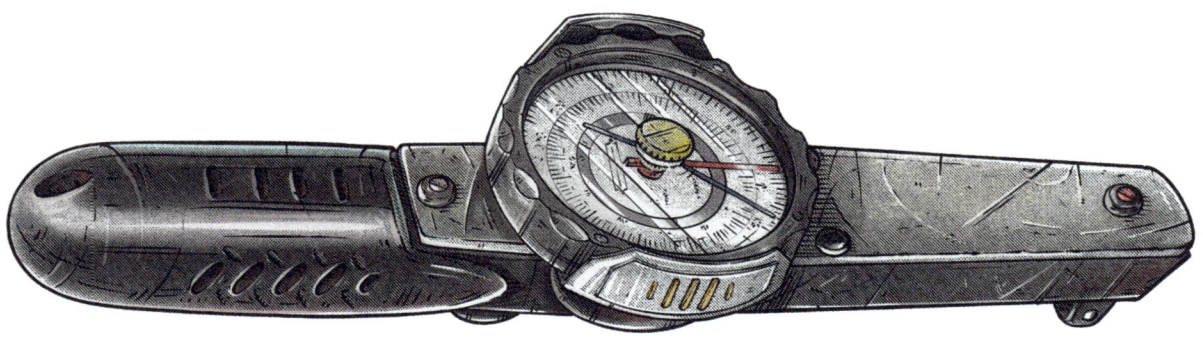

Tool No. 112

ALLEN KEY

I've included this tool because everyone uses them but hardly anyone knows who this Allen guy is (in the UK, anyway). An Allen key or Allen wrench is also known as a hexagon or hex key and is used to tighten and loosen bolts with hexagonal heads. The tool gained its popular name because it was William G. Allen, from West Hartford, Connecticut, who was awarded a patent in 1910 to form sockets in screw heads ("preferably hexagonal in cross-section", he wrote on the patent application) which could be turned by a key or crank.

They were being advertised soon afterwards as Allen Safety Set Screws "made from bar steel... and warranted as strong and effective as the best projecting screw made, besides eliminating danger from rotating shafts". They were strong, didn't stick out and were cheap to produce, which were all big plus points. Allen didn't stick around at the Allen Manufacturing Company – the company he co-founded in 1910. A few years later, he'd joined the Henry Wright Manufacturing Company (also in Hartford), which was well-known for making drill presses.

The hex key/Allen key is a tool that really hasn't changed very much, at least until 1964, when tool and die-maker John Bondhus started producing his Balldriver ball-end hex driver in Monticello, Minnesota. This added a rounded end to the hex key, enabling you to twist the tool at an angle so you can fasten and loosen difficult-to-access bolts and screws. Although, the contact area is so small that unless you're working with a brand new bolt that still has straight edges, you're going to struggle. But having said that, most of the time you will use a ball-end hex driver is with brand new bolts because they're both in furniture flatpack sets, so everyone has about 200 of them in their houses. On a related note, as of February 2021, IKEA joined forces with Japanese design studio Gelchop, which developed a

I've got a love/hate relationship with Allen keys. Well, it's mostly hate, if I'm honest.

lamp in the shape of a massive Allen key to praise this tool's contribution to the company's success.

So, full disclosure: I've got a love/hate relationship with Allen keys. Well, it's mostly hate, if I'm honest. If I come across anything held together with hex bolts on *The Repair Shop*, I feel instant dread because they round off so easily. The inside of the bolt is usually full of rust, paint or dirt, so you can't put the key in properly and are forever scraping out all of the gunk. On top of that, there are such small differences between the metric and imperial sets that the chances are, someone at some point will have ended up using the wrong ones and rounded the edges of the bolt or the key. So I use them as little as possible!

But, I do have a little nostalgic love for them too. That's because in my early teens I spent a lot of time on BMXs and mountain bikes, and they were always held together with hex bolts that you'd need a multitool equipped with little Allen keys to tighten and loosen. Back then, the bolts and keys were always stainless steel, so they were hard and didn't go rusty. And I was proud of my bikes, so I kept everything fairly clean. Well, as clean as you can keep a mountain bike and a BMX when you're a kid!

Allen keys

The Allen key has hardly altered in design since William G. Allen was awarded the patent in 1910 for creating hexagonal sockets in screw heads.

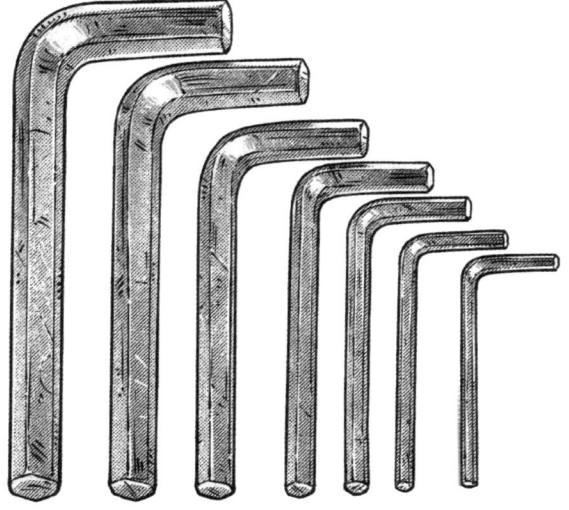

CHAPTER 8
SPECIALIST TOOLS

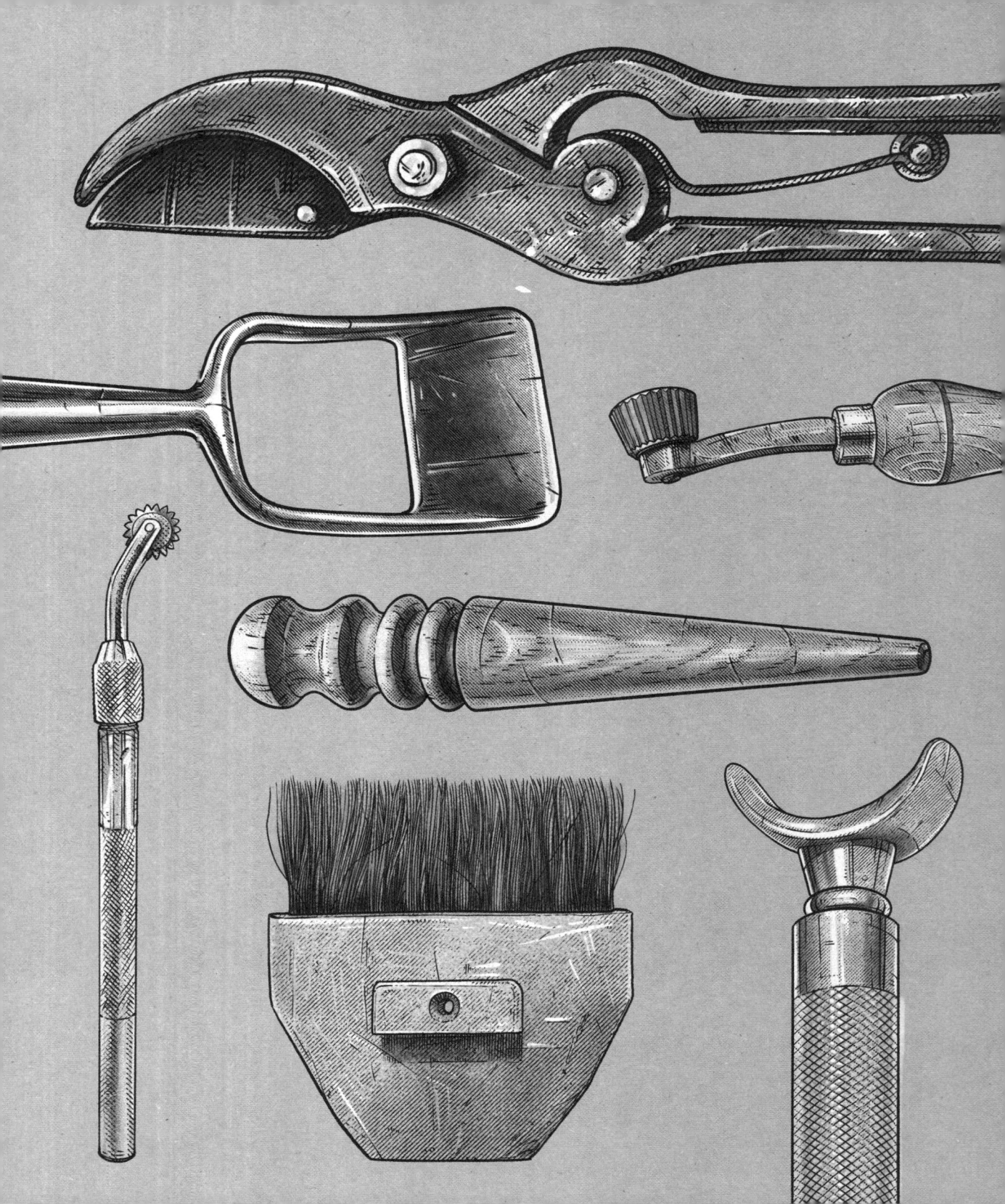

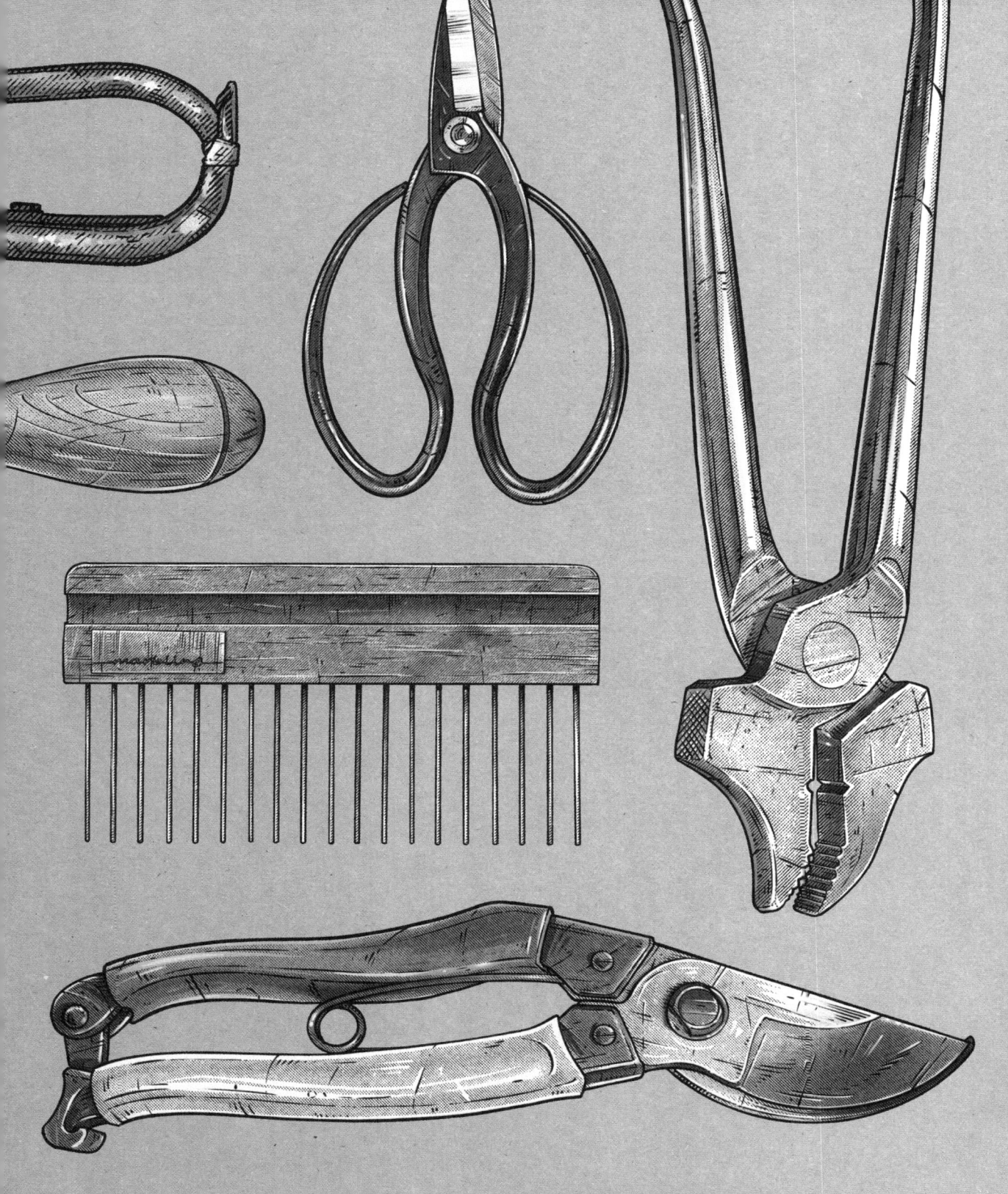

LEATHER BURNISHER

Also known as a slicker, this tool is used for smoothing and hardening the edges of cut and shaped leather so it will last and wear well. Burnishers are most commonly made of wood, but you can also get brass, glass and plastic ones. The earliest ones may well have been made from bone. Some burnishers have a number of grooves that correspond with various different thicknesses of leather.

To use a leather burnisher, you line up the leather with the groove that matches it and move the burnisher back and forth across the leather's edge, which generates friction. The heat produced joins together leather fibres and smooths them.

In 2013, archaeologists working at a cave in the Dordogne, in southwest France, discovered curved, thin implements made from the ribs of deer, which were more than likely used to generate friction to soften and toughen animal hides. What made this find really special, though, is that these burnishing tools, known as "lissoirs", were dated at between 40,000 and 51,000 years old and were discovered at a site used by Neanderthals. Until then, lissoirs had only been associated with early humans. This amazing find opened up the possibility that Neanderthals invented burnishers, because 40,000 years ago, modern humans hadn't arrived in Europe.

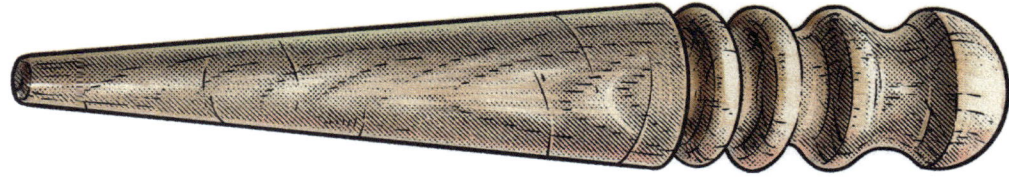

Tool No. 114

ROUND KNIFE

Also called a harness knife or a saddler's round knife, this is a versatile leatherworker's tool, which was in the past used to cut thick leather intended for saddles and harnesses. Everyone seems to have one at *The Repair Shop*, and I always think to myself: "that looks a lot like one of those fancy pizza knives, doesn't it?!"

One of the most famous round-knife producers is Vergez-Blanchard, a French toolmaker dating back to 1823, when Louis René Blanchard, a cutler by trade, set up a workshop in Paris. The company was awarded gold medals at the Universal Expositions in Paris in 1878 and 1889, and a coveted Grand Prix at the Paris Exposition in 1900, all for the quality of it's tools.

Another toolmaker who made fine round leather knives was Henry Sauerbier, although he was probably best known for the ornately decorated swords used by Union Army officers during the American Civil War. He was born in Germany in 1822, before emigrating to the USA in the late 1830s and settling in Newark, New Jersey. He took over a tool company founded by John H. Crawford in 1848 but started a new business under his own name in 1855, producing leather and shoemaker's tools. Five years later, the company was hit by a devastating fire that caused around $50,000 of damage (around £1.75 million today), but the business survived. Then, in 1876, an exploding boiler killed two of his workers and seriously injured both Henry and one of his sons. Henry died in 1886 and left the company in the hands of his sons.

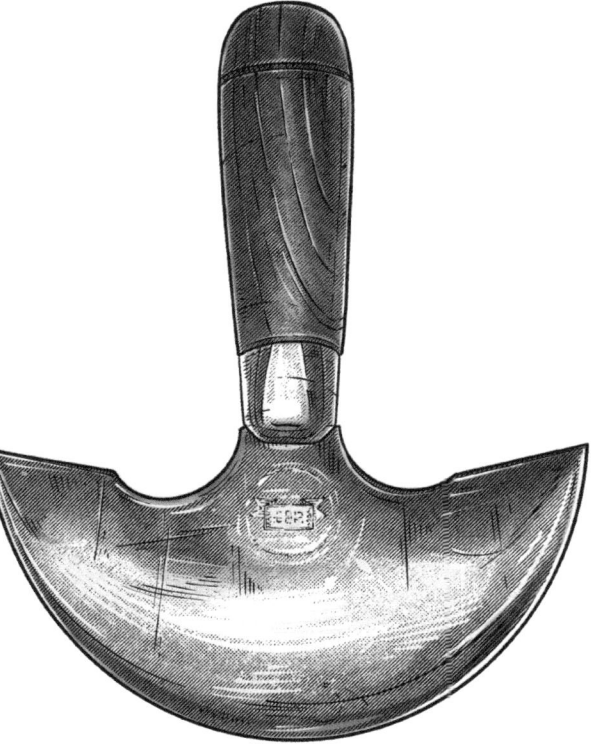

Tool No. 115

SWIVEL KNIFE

This leatherworking tool has a chisel blade and a barrel connected to a U-shaped "saddle", which is mounted on a pivot. It's designed for carving decorative designs into leather, and you use the knife by placing your index finger in the saddle, with your thumb and fingers gripping the barrel.

The swivel knife is often associated with intricate carved decorations on saddles in the USA. One of the most famous saddle-makers and leather toolers was Don King, born in Wyoming in 1923. He was the son and grandson of cowboys and started learning how to tool leather in his early teens. After serving with the US Coast Guard during the Second World War, Don settled in the town of Sheridan, Wyoming, and became an apprentice to an expert saddlemaker. He went out on his own a year later and set up a shop, but soon changed his focus, buying a 200-acre ranch and raising horses and cattle. It wasn't until 1957 that he became a full-time saddlemaker, and two years later, his designs were so well thought of that he was asked to design the saddles for the Rodeo Cowboys Association World Championship. King is credited with developing what became known as Sheridan Style – a type of saddle with its distinctive decorative wild roses in patterns of interlocking circles. King's saddles have travelled far and wide. Former US president Ronald Reagan has one, as does Queen Elizabeth II.

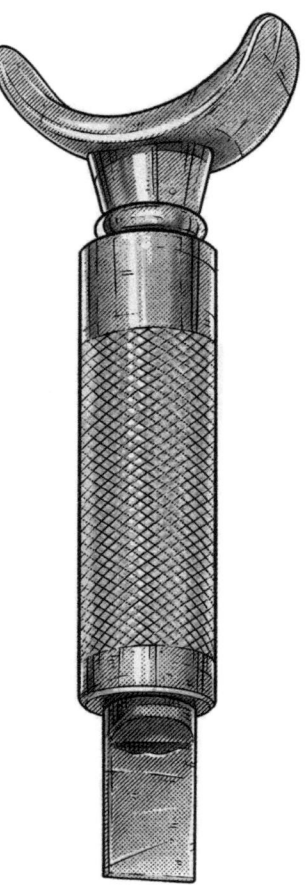

SIGNWRITING

I'd never learnt signwriting before the time I had to out of necessity when I was doing set design. I was doing a commercial job and the client wanted their logo on the front of something, I think it might have been the front of a bar. The go-to thing in the events world is a vinyl sticker – quick, easy, cheap – slap it on. But I didn't want to do that. For one, I didn't have a wide-format printer, but also, I wanted to learn the craft, and it would just cost me a bit of time. Plus, I could age and distress it so it looked old, which would look a lot nicer than a vinyl one.

I spent a couple of weeks with Joby Carter, the guy who runs Carter's Steam Fair, the largest vintage travelling funfair in the world. He's the most talented signwriter I've ever seen, and it was just brilliant getting to see him work and learning from him. That set me off, and before long everything I did was getting a pinstripe around it because I knew how to use a pinstriping brush.

I'm not a signwriter, and I don't like it when people call me one, because some people do it day in and day out freehand and they're incredible. I'm happy putting a logo on something once it's been printed out, so I can work from a template.

I bought three brushes from Joby, which I've got in my engineer's cabinet, in which I keep my signwriting brushes. The ones from Joby are sable brushes, which are highly prized among artists and signwriters. One of the most famous names in the signwriting world is A.S. Handover, who have been making world-renowned sable brushes for about 60 years in Welwyn Garden City, Hertfordshire.

LINING BRUSH

I've got a brush for lining, which has a very thin tip with very long bristles. They need to be that way to hold the volume of paint and help you keep a straight line. You wouldn't believe how long a line you can paint with one of these if you're really good at it.

I've bought a couple of synthetic brushes, but I just can't get on with them. The quality is a lot lower. My favourite brush is the chisel-edged brush, which, for whatever reason, is a little wonky, so the left-hand tip is slightly higher than the right-hand one, which gives it a nice point. It's also got a kink in it and it just fits with my hand. You can always tell the ones I use the most because they're covered in paint! Whenever I try and work with a new one, I always struggle because it's not quite the same. You get attached to them.

To clean my brushes, I use white spirit and then they get greased, normally with a solid grease or petroleum jelly. Just a tiny bit will hold them in shape. Be careful when you're carrying them, because they can get damaged if they're loose in a tin. It's best to get those holders that keep them still, so you don't find a tin full of bent brushes, which has happened to me a couple of times.

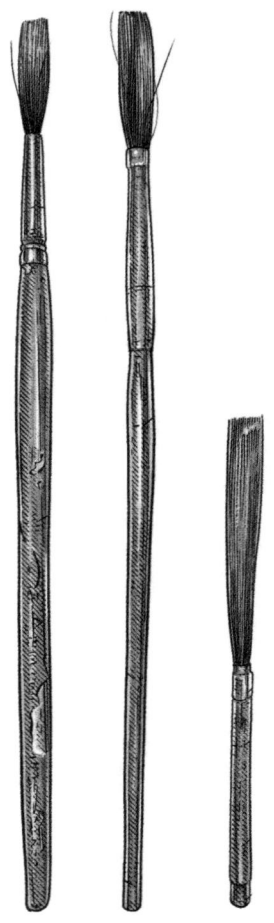

Tool No. 117

POUNCE WHEELS

A pounce wheel, also sometimes called a tracing wheel, is a tool with a little wheel at the end containing a certain number of teeth. You use it to trace patterns so you can transfer them from one surface to another. It's like making your own stencil. Pounce wheels are commonly used by signwriters, woodworkers, sewers and upholsterers. But they've been used for centuries to make copies of paintings.

To use a pounce wheel, first lay transparent or translucent paper over an original image, then track along all of the lines of the original image using the pounce wheel, which perforates the paper. You'll end up with a drawing made of pricked holes that you can lay onto a new working surface. Then when you've got the paper in the right position – where you're going to be drawing your sign, for example – rub a pounce pad over the top.

A pounce pad looks like a small board rubber that a teacher might use on a white board (well they did in my day, anyway!). The pad has a chamber on the reverse side that you fill up with pounce powder, a very fine powder usually made from chalk or charcoal. When you dab it or brush it all over the paper you perforated earlier, you'll end up with an outline of the sign. Now you're ready for the hard bit – painting the sign!

Pounce wheel

The Hertfordshire-based company A.S. Handover (see page 193) has been producing tools for signwriters and gilders, including pounce wheels, for over 60 years.

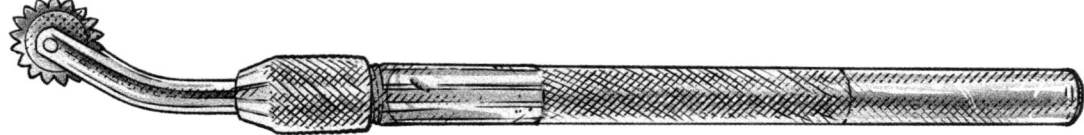

Tool No. 118

MAHLSTICK

A mahlstick usually has a metal or wooden rod with a rounded chamois leather or suede head. I've got a cheap metal one that comes in two parts that screw together, and a rubber head, but if I haven't got it with me, I really miss it! Like with the signwriting brushes (see page 193), I've got Joby Carter's voice in my head telling me to "use your mahlstick". It's tempting to just do without one, but it's one of those tools that experts have been using for centuries. You've got to trust that knowledge that's been passed down generations of pros who have been trained properly and have become among the best in the world at what they do. And that means shutting off the voice in your head that says, "but it'll be quicker to do it like this!"

It is a weird feeling using a mahlstick, it feels a bit unnatural and some people can't get used to it, but it's worth persevering. You'll thank me for it, as I've thanked my teachers. It's a perfect example of believing in the craft and not thinking "I know best".

Mahlsticks are used when painting and signwriting in the hand you're not painting with, to balance and steady the brush hand. It's also really helpful to stop you from accidentally smudging what you've already done, which happens a lot because you naturally want to hold on to the surface. That's why a mahlstick has a non-marking suede or leather head. It's quite a long stick, usually 1m (3ft) long. I also find that using one allows for much freer movement with my painting hand. By pivoting one hand and adjusting the other to suit, you can draw a line parallel to another line you're painting by using both hands, giving you a bit of control.

Mahl comes from the German *malen*, which means to paint, and mahlsticks have been around since the Renaissance. They usually come in two parts, which you screw together like a snooker cue, and these days you can get some fancy ones that are made from super-lightweight carbon fibre.

Mahlstick

Mahlsticks were traditionally made from a two-part wooden rod with a rounded chamois leather or suede head, a design that has hardly changed over the years.

Mahl comes from the German "malen", which means to paint, and mahlsticks have been around since the Renaissance.

COOPERING

It's quite an art, making a cask, and it takes several years working as an apprentice before you can call yourself a master cooper. Fitting the hoops to a cask was actually a job in its own right, undertaken by a hooper, a kind of assistant to the cooper. Before the 19th century, these hoops would have most commonly been made from flexible woods such as willow or hazel and were known as "withies", but now they're made of galvanized iron or steel.

Sun plane
This specialized coopering plane (also called a topping plane) has a curved edge and is similar in design to a jack plane (see page 152).

Chiv

A chiv is used to scallop out and level the inside of a barrel.

A traditional cooper uses a number of specialized tools, and some tools we've already talked about, to make wooden casks. Here's how the process goes. First, they use a side axe (a short, one-handed axe with a long cutting edge offset from the handle) to cut wood into staves, then they use a draw knife (see page 84) to hollow, taper and bevel the staves, before smoothing them with jointer plane (see page 155). Next, they insert the staves into the raising hoop (the smallest of the metal rings around the container). Now begins the process of steaming the wood so the staves can be bent into shape without splitting, and more hoops are fitted using a hammer. The ends of the staves are shaped with an adze (see page 117) to give them a bevelled edge and then levelled and smoothed using a sun or topping plane; a specialized plane with a curved planing block but not dissimilar to a jack plane (see page 152). Then a chiv, a planing tool with a projecting steel blade, is used to scallop out and level an area of the interior of the barrel before a circular planing tool called a croze cuts a groove onto which the heads (both top and bottom) of the cask will sit. Next the hoops are fitted to each end and the interior of the cask is smoothed down. The heads are made and shaved down using a bow saw (see page 102), then the edges are shaped to fit it in the croze (groove). Once the outside of the cask has been smoothed and both heads fitted, a hoop driver is used (a short, wedge-shaped tool) with a hammer to force the hoops into place. Lastly, an auger, a tool that works like a corkscrew, drills into the side or top of the cask to make the bung-hole. Phew!

There are very few master coopers around anymore in England, but it's a different story in Scotland, because of the number of casks made and used to mature Scotch. But one of the remaining English master coopers, who has turned into a success story recently, is Alastair Simms. In 2020 Alastair was staring down the barrel of unemployment until Christian Jensen, the founder of Jensen's distillery in London, offered him a job setting up Jensen's Cooperage, making and repairing

Croze

A croze is a circular planing tool used to cut a groove into the barrel onto which the heads of the cask will sit.

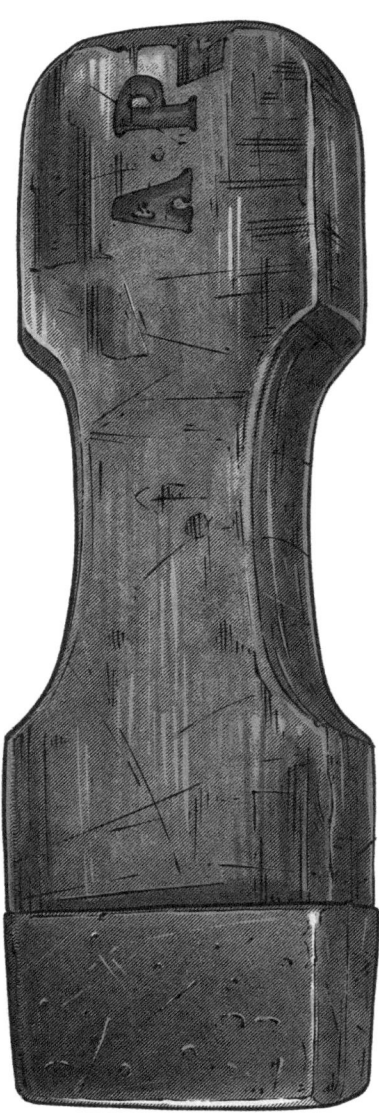

Hoop driver

Once both heads have been fitted, a hoop driver is used with a hammer to force the hoops securely into place.

barrels. With the booming English wine industry and appearance of more whisky distilleries, Alastair's been joined by two more master coopers. Life has been breathed into the craft once more.

I'd actually never met a cooper (and didn't know that much about what they did) before Alastair came along to *The Repair Shop* to fix a knackered butter churn. He made such a lasting impression here that they added a clip of him to the title sequence (which stayed for years), in which he tells me: "I'm going to teach you how to do it the old-fashioned way." I'd suggested to him that we cut a piece of hoop iron using a grinder and he wasn't having any of it!

Alastair and I worked on the butter churn together, and it was a pleasure getting to know him, watch him work and basically become his apprentice for a bit. I still can't believe how heavy coopers' hammers are – I think the one he used was around 2.5kg (5½lbs). I remember when he shook my hand, I thought "his forearm's about the size of my leg!"

The cask was in a bad way, leaking water in about ten different places, but Alastair was completely confident we'd fix it up no problem. He's the kind of guy that loves what he does, knows exactly what he's doing and always has a smile on his face, so he's the perfect teacher. He also knows how to take the mickey out of you, which meant the whole job was a lot of fun. I feel very lucky to have learnt a bit about coopering from the best in the business.

GILDING

Gold leaf is made by melting gold in a crucible before pouring it into a mould. After that, it's fed through a rolling mill, then the space between the rollers is adjusted so that the gold thins. This process is repeated until the gold is thin enough to begin hammering, or goldbeating. Goldbeating is relentless work, and involves striking the gold thousands of times until it's about 0.1 micrometres thick. To put that in a way that is easier to imagine, that's hundreds of times thinner than a human hair.

The product you end up with is gold leaf, which is cut into squares and (nowadays) packaged up in little booklets, with each leaf separated with tissue paper. Gold leaf comes in a number of different shades and carat values, from pure gold (24 carat), which won't tarnish or darken over time, to 12 carat (white gold) which is normally 50 per cent gold and 50 per cent silver. The top-quality stuff is incredibly durable and easily outlasts paints and vinyl, because it doesn't rust, doesn't suffer in bad weather and is resistant to most chemicals.

The layering of gold leaf over an object or surface is called gilding. There are two different ways of going about it: water gilding and oil gilding. Oil gilding involves using an oil-based adhesive to make the gold leaf stick to the surface. Water gilding is a much trickier technique and involves many stages, but its name comes from the fact that water is involved in the preparation of the adhesive and in the application of the leaf itself. Water gilding produces the best, longest-lasting results. For that reason, it's seen as the summit of the craft.

Working with gold is a specialist skill. I did one of David Smith's gold-leaf workshops down in Devon, where he teaches oil gilding, water gilding and reverse glass gilding – the technique you see used on old pub signs, windows and mirrors. I learnt so much and have put those skills into practice many times since.

GILDER'S KNIFE

Made either from carbon steel or stainless steel, if you've got something in the bank for it, this knife is designed to cut and trim gold leaf, but you can also use it to pick up and manipulate gold leaf.

The blade of a gilder's knife is usually about 15cm (6in) long and attached to a wooden handle. You cut gold leaf on a gilder's cushion (also called a gilder's pad), which is usually made of a soft material such as chamois leather. Some gilder's cushions even come with a little windshield to protect gold leaf from draughts. Gold leaf is so light that it'll blow away if you're not watching what you're doing.

Gold leaf comes in sheets, rolls or little booklets, which is how I usually get it. It's available in all sorts of carat values and shades, from 24 carat (pure gold) to 12 carat (50 per cent gold and 50 per cent silver). You need to be careful not to touch it with your hands because it's so fragile and it'll stick to you. Which is why you need a gilder's tip (see page 204) to pick it up.

Gilder's knife
These specialist knives are used to trim and shape gold leaf on a gilder's cushion.

Tool No. **124**

GILDER'S TIP

My gilder's tip is made of two bits of cardboard glued together to hold the bristles in place and a handle made from the little bit of plastic that holds kitchen worktops together.

David Smith (see page 202) put the worktop handle on as a makeshift measure, but it works perfectly and I've been using it for years. Everything else I've got for gilding is quite posh-looking, expensive and made by the best in the business, but I can't use another gilder's tip now. I like that Dave did it, and I had such a good time on the course with him, that I'm reminded of it whenever I use it.

To use a gilder's tip, open up your gold-leaf book after making sure you haven't got a draught in the room, or else it'll flap about. One trick I learnt from David was to put a little bit of grease on your forearm, dab the brush onto it and it'll help you pick up and apply the gold leaf. When you're water gilding, you'll have the gilder's tip in your dominant hand and the gilder's mop in the other, so it's pretty labour intensive.

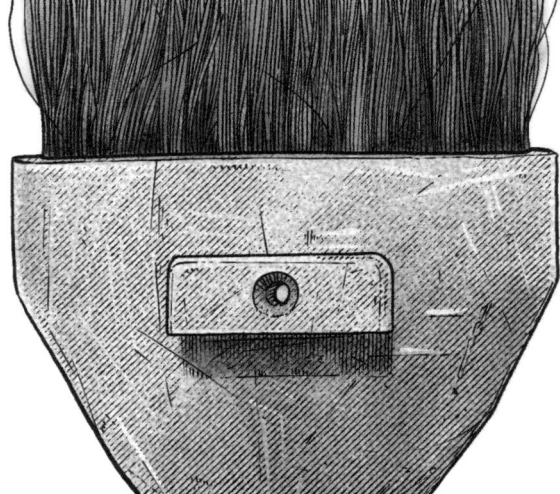

Gilder's tip
My handmade gilder's tip, featuring a kitchen worktop handle.

Tool No. 125

GILDER'S MOP

Once you've put the loose gold leaf on to whatever you're gilding (see opposite), you need to use a brush known as a gilder's mop to help settle it into position and brush away the loose bits. It's a domed brush (it looks a lot like a makeup brush) and is usually made with squirrel or pony hair, which won't scratch or dull the gold.

You tend to see more expensive gilder's mops around with a quill, secured with several rings of copper or brass wire to hold the base of the bristles together, rather than a metal fastening. In the past, all painter's and gilder's brushes would have been quill-bound, made using real bird quills (the hollow tube that runs down the centre of a bird's feather). These days, they're all synthetic. But they *look* old and crafty, which is one reason why a lot of people like them. And I'm one of them, if I'm honest!

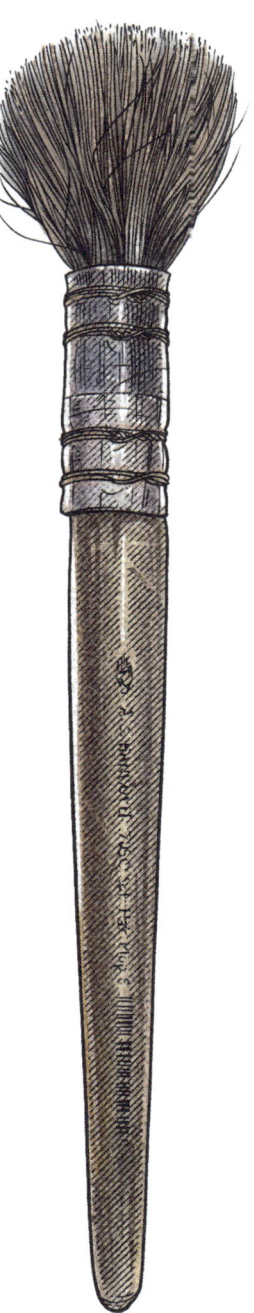

Gilder's mop

These soft brushes are used to gently remove any excess gold leaf, known as skewings.

GARDENING

I really wanted to keep bees, partly because I knew they were in trouble, but also to help pollinate the shrubs in my garden. I'd never had anything to do with a hive or met anyone who had, but I was intrigued. The more books I read, the more I realized that there are a million different ways to do things when it comes to beekeeping. The rules are completely different to fixing stuff on *The Repair Shop* because you're dealing with animals, so some things are completely beyond your control! I was getting overwhelmed with all the different approaches and opinions, so I figured let's just get them and see how we go. Sometimes learning on the job really is the best way.

Having the bees has made us more aware of what's in the garden, because at some times of the year, such as November, we realized that nothing's in flower. And that means you've got some hungry bees having to make longer trips elsewhere to keep them going. So it makes you garden in a different way, thinking carefully about what you plant and where. Bees always struggle in the autumn after most of the flowers and plants have gone over and they're looking for as much energy as possible to keep the hive going during the winter, when they hibernate.

Tool No. 126

BONSAI SCISSORS

With their large handle loops and delicately pointed blades, these scissors are perfectly suited to the fine art of trimming and pruning bonsai trees. They're also beautiful tools in their own right, and the most incredible ones I've ever seen are made by Yasuhiro Hirakawa, who is quite possibly the only traditional scissormaker left in Japan.

Yasuhiro Hirakawa is a fifth-generation master craftsman with over 50 years of experience, and he's an absolute legend in Japan, although he's a humble man and only acknowledges that he's just beginning to reach his peak. He sells his creations – kitchen knives, ikebana clippers (ikebana being the Japanese art of flower arrangement) and garden shears, as well as bonsai scissors – under the brand name Sasuke and works in Sakai, a city just south of Osaka with a strong history of blade-making.

Not only does Hirakawa produce super-sharp, extremely well-made bonsai scissors, he also decorates them using beautiful and highly intricate gold and silver motifs. Sasuke's bonsai scissors are made to order, so some of the special requests he gets can take him up to a year to research and make, such is his commitment to his craft. The most expensive pair he has ever sold was ¥3,600,000, which isn't far off £25,000, but most seem to go for more like £800. But then, bonsai is an art form and you need the perfect pair for the job!

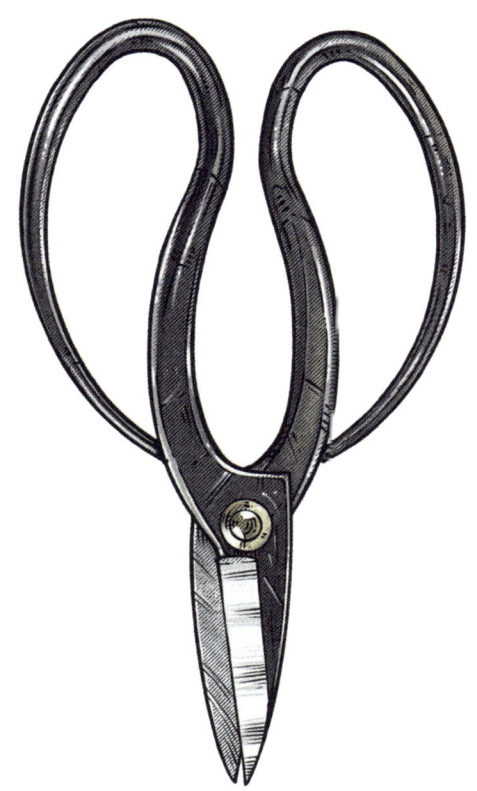

DIBBLE

An amazingly simple gardening tool, the dibble (or dibber, as it's known in North America and parts of Europe) is basically a sharpened stick with a tapered point.

A dibble is used to poke holes into soil so you can plant seeds, seedlings and bulbs. The design of the dibble has changed very little since Roman times, and, to be fair, it doesn't need to, because it does the job. There are a few modern varieties of the dibble, though. The traditional dibble has a curved handle, but you can also get one with a T-shaped handle, which lets you apply even pressure so the shape and depth of the holes stays more consistent.

From the 18th century, long-handled dibbles were first developed, and they were often used by two-person teams – one digging the hole and the other filling the holes with seeds. In the 19th century, the first stainless-steel dibbles were made, and this is what the tips of most dibbles are made of today, attached to a good weather-resistant hardwood handle.

Many of the grass areas at Great Dixter (one of my favourite gardens and also close to my workshop) are planted with crocuses, snowdrops, fritillaries and daffodils, so in February and March, they are just a sea of colour. It's inspired me to buy 200 bulbs and plant them underneath my lawn in autumn 2022 so the following spring I have a little bit of Great Dixter at home.

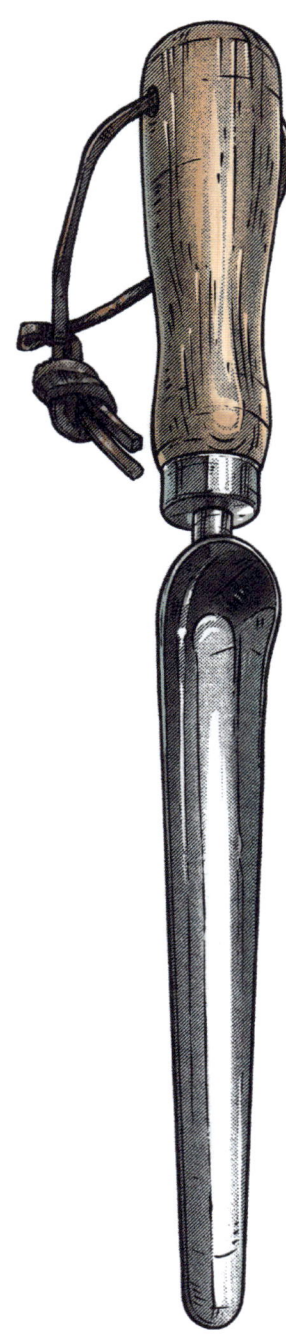

WIDGER

A widger is a cross between and dibble and a trowel. It has a narrow, scooped head and is used mainly for separating, transplanting and planting seedlings and seeds.

It's also handy for weeding, especially in tight spaces. Vita Sackville-West, the 20th century author and garden designer who co-created Sissinghurst Castle Garden (one of the most amazing gardens in the UK, if not the world, and just down the road from my workshop, so I go there quite a lot) with her husband Harold Nicholson, had a soft spot for the widger, calling it "the neatest, slimmest and cheapest of all gadgets to carry in the pocket".

Vita tells us that garden writer Clarence Elliott (1881–1969) invented the widger and praised him for coming up with the word, because its name actually reminds you of the action (and maybe even the sound) of prising up a weed. Vita even went on to say that the widger is useful for all sorts of other things, such as serving as a letter-opener, screwdriver and toothpick. She deemed it "the perfect gadget", and that's high praise!

Tool No. 129

TICKLER

The original tickler (although it wouldn't have been called that at the time) was made by a fisherman in north Wales who used it for spearing fish. And when you look at it, it screams *old sea dog*. I mean, it looks like Poseidon's trident, for one!

So the story goes that this fisherman gave his tool to Fergus Garrett, the Head Gardener at Great Dixter in East Sussex, one of the most famous gardens in the country. Fergus started using it in the garden, and passed it around to the other gardeners, who loved it. Great Dixter is a cottage garden where the borders are packed with plants, so you need something precise to get into the tiny spaces between plants to weed, add compost or plant new bulbs. Fergus made a couple of adaptations to make it even more useful in the garden, such as adding a longer handle so you don't have to stand in the flower beds themselves or lean over when you're using it. He also extended the central tine to single out weeds. And so the tickler, or tickling fork, was born.

A couple of years ago, my wife and I wanted a change in pace, so we sold our London flat and looked to buy a place in the country with a bit of space. We ended up falling in love with an old Georgian cottage with a mature garden that used to be opened up a few times a year as a show garden. We love the place, although whenever there's a storm, we wince, because we know some of the old peg tiles might be slipping off the roof. The elderly couple who moved out couldn't keep up with the maintenance of what is a really special garden, so my wife and I decided to get stuck in, despite never living in a place with a garden like it before. Since then, we've really got into it – I've even become a beekeeper! I feel like we're custodians of the house and garden, and it's our responsibility to look after it properly and keep it going. Plus, everyone in the village is always asking about it. It's funny that I seem to have taken what I do for a living at *The Repair Shop* back home with me!

While I was slowly becoming green-fingered, I went to Great Dixter one day to look at the gardens. I loved it, particularly how much of the inner workings you can see as a visitor. So I ended up booking tickets to some of Fergus's lectures about how to look after a garden. He told us about the tickler, and he was preaching to the converted, because I love a new, obscure tool. But more than that, I love a tool for a job that I didn't even know existed. Now, every time I take it out of the shed, I think of Fergus and Great Dixter. It reminds me of long summer days learning new things with amazing people.

Tickler

Netherlands-based company Sneeboer have been making hand-forged garden tools, including ticklers, since 1913.

Tool No. 130

HOE

Burgon & Ball, which sounds like a shop you might find in Diagon Alley from the *Harry Potter* books, has been going in Sheffield since 1730. It first made agricultural shears, which it became famous for, but then started making gardening tools in the 19th century, and by the 1920s, this had become the biggest part of its business.

In the 1922 catalogue, there were 96 different types of hoe! Any tool that has 96 different variations in a trade catalogue must be worth spending some time on. And it sure is, because the hoe is a truly ancient tool that once was essential for survival, being that its main functions are to shape the soil to plant seeds and bulbs, to clear weeds and to harvest crops.

There are four main types of garden hoe – the Dutch hoe, draw hoe, stirrup hoe (also called the stirrup or oscillating hoe) and ridging hoe. The Dutch hoe is the one that we're probably most familiar with, with its D-shaped head and sharp, wide blade. You use it standing upright, pushing or pulling the head through the surface of soil to remove weeds. A draw hoe is a simpler-looking tool, with a rectangular paddle-like blade set at around 90° to the angle of the handle. You use it to chop into the ground before pulling the blade towards you to sever weeds from their roots and to loosen and aerate the soil. The blade is good for cutting into and breaking up thick, compacted chunks of earth.

The stirrup hoe, as you might expect, has a pivoting head that looks like a horse's stirrup. The blades cut in two directions so you can use it really effectively in a push and pull motion to get rid of stronger weeds. A ridging hoe is a triangular (or sometimes heart-shaped) hoe that's great for digging trenches for planting seeds and also for breaking up hard soil.

There are four main types of garden hoe – the Dutch hoe, draw hoe, stirrup hoe and ridging hoe.

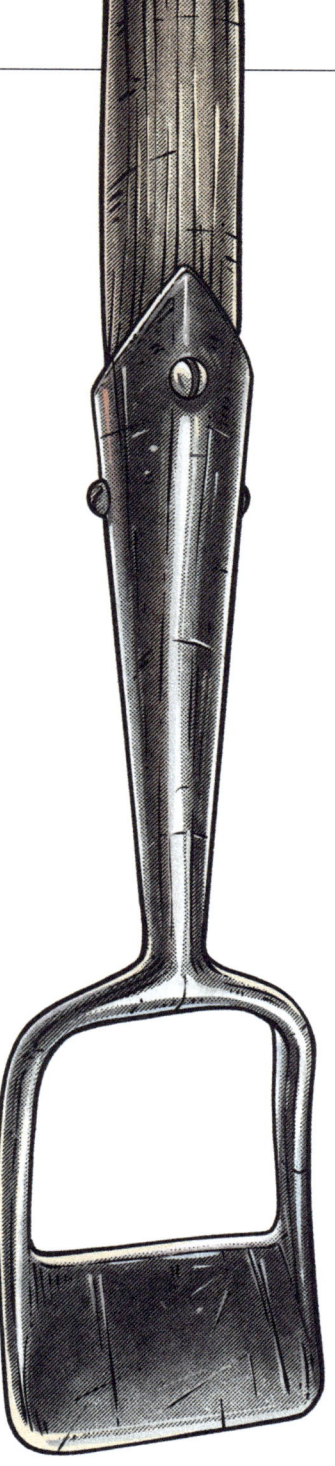

Dutch hoe

This is the hoe we're probably all most familar with. It's easy to use and has a wide blade designed to reach just below the soil's surface.

PRUNING SHEARS

Pruning shears (also known as secateurs in the UK) are specialized scissors with curved blades used for cutting plants. There are two main types. The first are bypass secateurs, which anyone who's ever done an afternoon of gardening will be familiar with. The cutting action works like regular scissors, with one blade crossing the other to perform the cut. Anvil pruners are the second type, and they work with more of a crushing action than a cutting one, with a sharp blade forced downwards to a flat metal base, which is softer than the blade. This works much like a hammer and anvil, hence the name. They're often used to clear dead wood because their cutting method can cause damage to live plants. Bypass secateurs aren't as powerful but create a cleaner cut, which is much better for the plant. Just make sure to keep them sharp.

The supposed inventor of secateurs was French aristocrat Antoine-François Bertrand de Molleville (1744–1818). Well, he's the man who gets credited with the invention in the French horticultural almanac *Le Bon Jardinier*. But it seems it took a while for people to warm to them. Apparently in 1840, secateurs caused a serious problem in the town of Béziers in southern France.

The story goes that the town's agricultural committee met to discuss whether or not this new-fangled tool was better than a traditional pruning knife for cutting vines. Local vine-dressers (who pruned and cultivated vines) got wind of this meeting and reacted very angrily, worried that secateurs would mean they'd lose their jobs. So 300 of them took to the streets to voice their anger. It was clearly quite something to behold as a newspaper article referred to it as a riot. They eventually took hold in France, but didn't arrive in the UK for some years, and they weren't warmly received.

In the 1871 book *The Horticulturalist* by famous horticulturalist John Loudon (now revised and edited by well-known gardener and author William Robinson), he has to persuade readers to give secateurs a go: "The sécateur is a French instrument that every gardener should possess himself of. I know well the prejudice that exists in England among horticulturalists against things of this kind, and their almost superstitious regard for a good knife… but when I saw how useful the sécateur is to the fruit-growers of France and how easily and effectively they cut with it exactly as desired, I become at once converted."

One of the most renowned names associated with pruning shears is Okatsune, a Japanese company based in Hiroshima that was founded by Tsuneichi Okana in the late 1940s. Its pruning shears are well-known for being super sharp, light and strong, which is everything you need in a pair of secateurs. They're also bright red and white, which is useful when you have a habit of leaving them somewhere in the garden while you get a cup of tea.

Fergus, the head gardener at Great Dixter (see page 208), told me that he never goes out into his garden without his shears, and come to think of it, all the gardeners there have shears in leather pouches on their belts. That's good advice, because whenever I'm out in my garden, there's always something to prune, chop or deadhead.

Pruning shears

The French secateaurs (top) may look a little different from the Japanese (bottom), but they perform the same fuction.

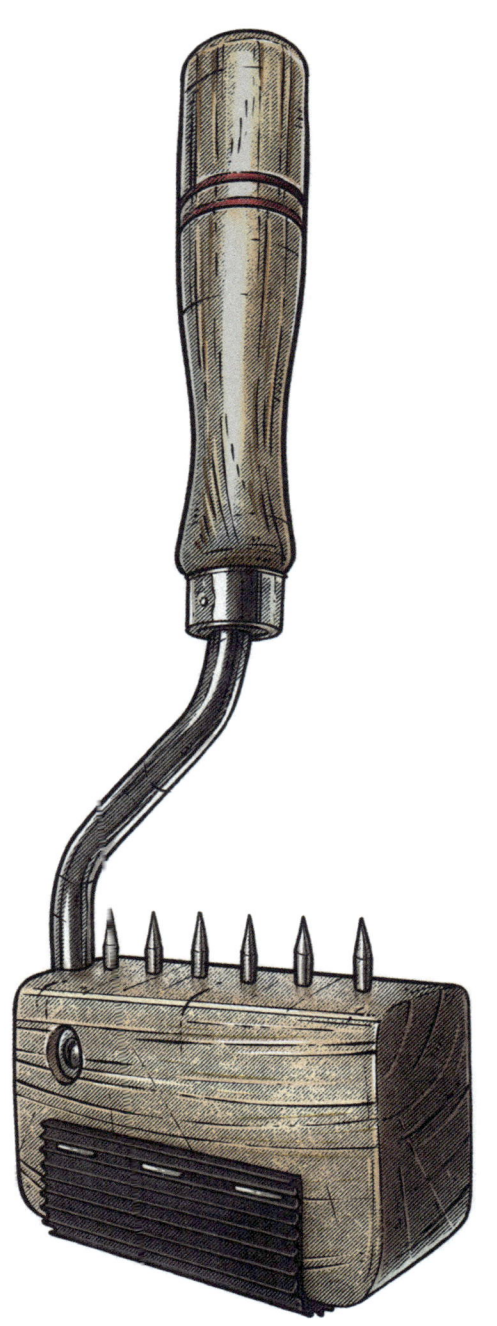
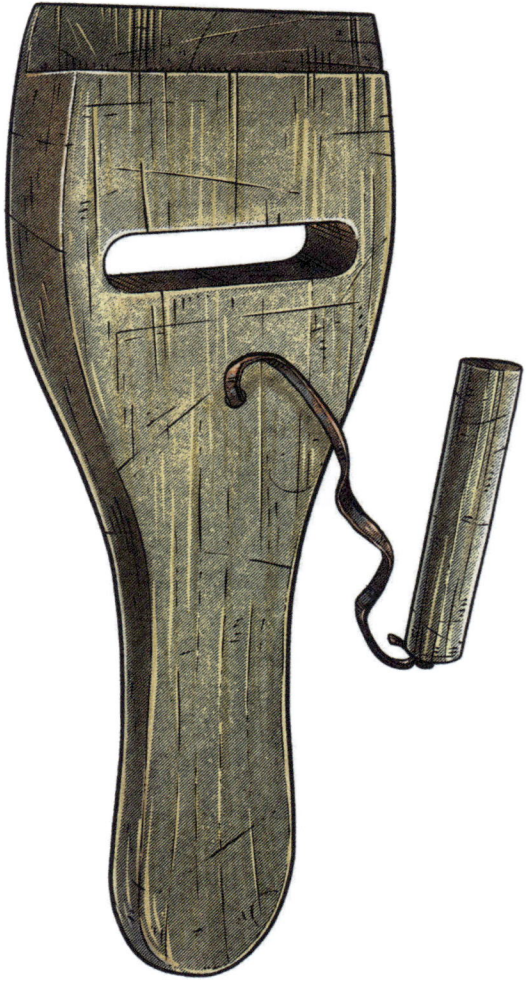

Webbing stretchers
The gooseneck webbing stretcher (left) and the more simple webbing stretcher (above).

Tool Nos. 133–134

WEBBING STRETCHER & GOOSENECK WEBBING STRETCHER

A webbing stretcher is a simple wooden tool used to stretch webbing (strong, closely woven fabric produced in a roll) so that it lies taut across the frame of the item of furniture you're upholstering.

It's a simple wooden bat-like tool with an oval-shaped opening, a groove and a dowel pin attached by a chain. It's a staple tool for a traditional upholsterer, like a hammer for a panel beater, and the first time I saw someone using one was actually Jay Blades at *The Repair Shop*.

First, you hammer tacks into one end of your webbing to secure it, then you stretch out your webbing, grab your webbing stretcher with the groove facing towards the furniture frame and thread it through the opening. Then you secure it with the dowel and chain and lever the webbing stretcher towards you. You're basically trying to trap the webbing between the frame and the webbing stretcher. Lastly, you hammer your tacks in to the webbing to secure it.

There are two other types of webbing stretcher – a gooseneck webbing stretcher and a spiked webbing stretcher. The gooseneck webbing stretcher has several sharp spikes at the base of its head to hold the end of the webbing, and a rubber shield on the top of its head to protect the furniture frame. Then you hammer in the tacks or staple the webbing to secure it. The spiked webbing stretcher is a similar tool to the gooseneck webbing stretcher but is cheaper, smaller and does not give you as much leverage.

A webbing stretcher is a simple wooden tool used to stretch webbing so that it lies taut across the frame of the item of furniture you're upholstering.

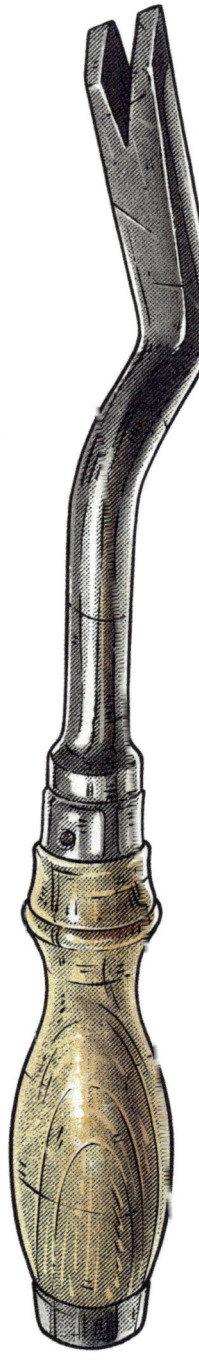

RIPPING CHISEL

Used to lift off the old fabric that you need to remove before re-upholstering, ripping chisels come in a few different forms. There are straight ripping chisels and bent ripping chisels, both of which have either a conventional chisel blade or a V-shaped claw, so you can lever out nails and tacks. I use the V-shaped version quite a bit for taking out door-panel clips in classic cars, such as old VW Beetles. It's the only thing that can really get behind the door cover and pop the clips out. It's handy for so many little jobs like that. I think of a ripping chisel as a smaller version of a claw hammer (see page 63) or a pry bar but a lot more gentle and much more manoeuvrable with its size and shape.

Two of the older upholstery suppliers were set up in the 19th century on either side of the Atlantic: C.S. Osborne & Co, founded in 1826, in Newark, New Jersey, and Glover Bros, established in 1899 in south London. C.S. Osborne makes specialized hand tools for upholsters, leather workers and a number of other trades. Founded by Charles Samuel Osborne, and still owned and run by the same family to this day, they do everything themselves, from forging the steel to sharpening the blades ready for use.

Glover Bros was set up by Thomas Glover Waine as the wholesale arm of a furniture, upholstery, carpets and drapery business called House of Waine, which opened its doors near Elephant and Castle in 1848. The retail shop was hit by a bomb during the Second World War and did not survive much longer, but Glover Bros continued, moving to the East End before relocating to Devon and lastly Somerset in 1985, where they remain today.

SHOEMAKING

I've learned much more about shoemaking in the past couple of years after meeting and working alongside Dean Westmoreland, a fantastically skilled shoemaker and restorer (and a lovely guy too), who joined us at *The Repair Shop* in 2021. He got in touch with me on Instagram in May 2020 after I put up a picture of my Redwing work boots (a classic US brand going for over 120 years and still based in Minnesota) that I'd had for seven years and completely trashed. In the post, I said that I'd taken them to a couple of shoe repairers but everyone said they were too far gone to fix, so I had to bite the bullet and buy a new pair. Well, Dean's an authorized Redwing dealer, one of the biggest in the country, in fact, and he sent me a DM saying, "Send them up to me – I bet I can fix them." He had such a positive, can-do attitude that I mentioned him to the producers on the show, and a few weeks later, he was on set working on a pair of running shoes that helped British athlete Audrey Brown win a silver medal at the 1936 Olympics in Berlin. I got to work with Dean on the job, cleaning up the metal running spikes. He's got an incredible set of tools and a beautiful old Singer sewing machine. He really knows his stuff.

Tool No. **136**

CLICKER'S KNIFE

A clicker is a craftsperson who cuts the different pieces of leather that make up a shoe. When I say it like that, it sounds easy, but it's a highly skilled job that needs expert knowledge of the thickness, grain, shade, marking and quality of the material you're working with.

They earned the name "clicker" from the sound their knives made as they struck the brass edges that protected the patterns that would be overlaid on to the skin of the shoe. You can get either straight-bladed or curved clicker's knives, the latter being the one you'll use for cutting corners. One famous design you might see is a traditional French clicker's knife marked L'Indispensable. It has a long brass handle and a blade that runs beyond the top of the handle, so you can keep sharpening (and resharpening) the blade, which clickers will do after each use.

Clickers earned their name from the sound their knives made as they struck the brass edges that protected the patterns.

Tool No. 137

LASTING PLIERS

This is a clever tool that's really two tools in one – pincers and an attached hammer face – allowing the shoemaker to do two jobs with one piece of kit.

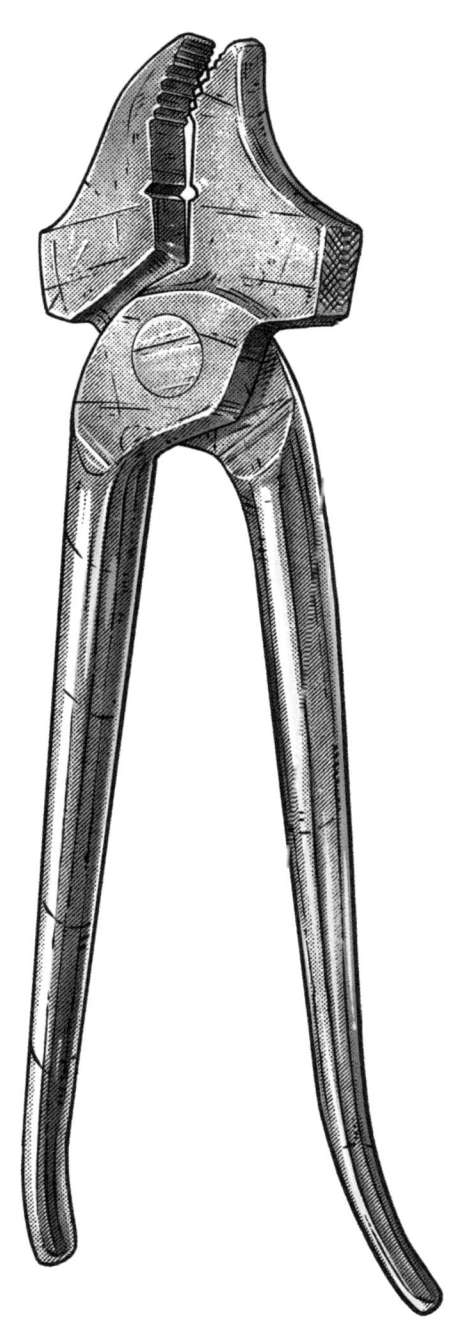

You can get lasting pliers with a square-ish looking head or a sharply curved head, but both feature pincers with serrated teeth that allow you to grip onto the shoe leather so you can manoeuvre it on to the lasting board (the paper-fibre board used when fixing the upper shoe material to the last, or mould). After the shoemaker has done that, they can use the convenient hammer face on the side of the head to whack in the nails or tacks, to secure the leather in place. Because lasting pliers can do two jobs without having to switch tools, they're often a vital part of a shoemaker's toolkit. Saving 10 seconds here and there really adds up over time, believe me!

Lasting pliers allow shoemakers to do two jobs without switching tools.

Tool No. 138

FUDGE WHEEL

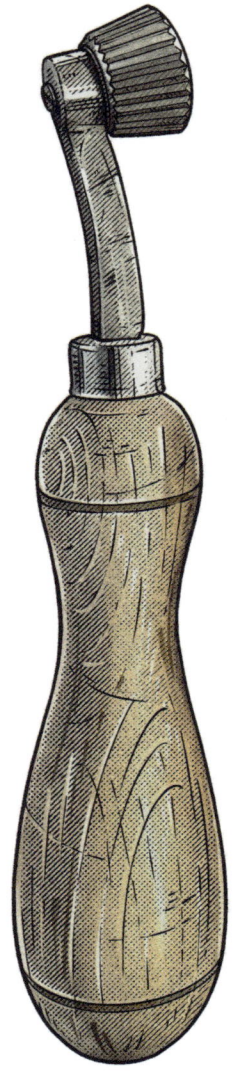

A contender for best-named tool ever, a fudge wheel is used for decorating the edge of a shoe's welt (the tiny piece of leather between the front of the shoe and the sole), to imitate stitching.

....

In the past, shoemakers would have marked out where the stitches would land on the welt before stitching the welt to the sole, but nowadays an adhesive is usually used. But the detailed stitching effect looks really smart, so it's added in. And it's this illusion that is the reason for the tool's name, because shoemakers "fudge" the number of stitches, with "fudge" being an 18th century word that basically meant fake or put together something slightly dishonestly. A fudge wheel comes in various sizes that are measured based on how many stitches it'll give you per inch. This measurement's usually abbreviated as SPI and the most popular are 8, 10 and 12, but the finest fudging wheels go up to 20 SPI. You warm up a fudge wheel on a little paraffin burner before you use it, so the wheel turns nicely.

A fudge wheel comes in various sizes that are measured based on how many stitches it'll give you per inch.

Tool No. 139
EDGING IRON

An edging iron sets the edges of the leather sole to give them an even, straight surface. The tool is heated and then pressed firmly along the edges of the sole, which strengthens it and stops water from getting in. This is important because these areas of the shoe get battered when you're walking, especially when the weather's terrible. They come in as many different sizes as you can imagine, because when shoes were all bespoke-made, you'd need a specific edging iron to suit the shoe you were working on.

Edging iron
Traditionally, edging irons were heated over a paraffin burner.

Tool No. 140

BONEFOLDER

A tool that sounds a lot like wrestling move, a bonefolder is a flat tool used to fold, crease and score paper, card, cloth and leather.

As you might expect, bonefolders are traditionally made of bone (usually from a cow or deer's leg bone, that been polished) although you can also get metal, wood and synthetic versions. They range from 12.5cm (5in) to 23cm (9in) long and come in a whole range of shapes, but the most common have either a pointed tip, which is handy for scoring, or rounded ends, which are good for pressing and burnishing.

Bonefolders are vital tools for bookbinders, who need to ensure the paper they are working with is as tightly creased as possible, but they're also used by paper crafters and leatherworkers, among other crafts. You use one by holding it between your thumb and index finger at a 45° angle to the paper or card and pushing down from the long edge of the tool.

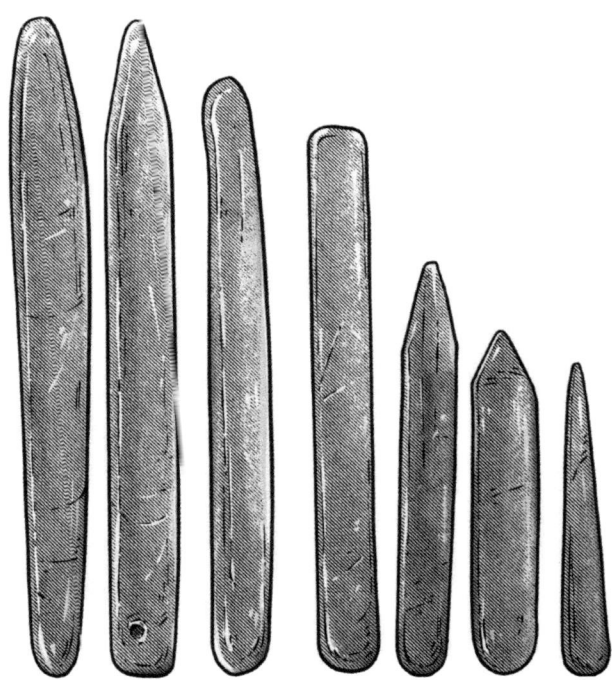

Bonefolder
As the name suggests, bonefolders were once made from bone. However, metal, wood and plastic versions are now available.

MARBLING COMB

Marbling is one of the things I remember doing for the first time in art at school when I was really young, and I remember being amazed by what you can create with just a few materials.

Marbling basically involves dispersing paints on top of a liquid (often a thickened water solution, called "size") and then capturing the effect on paper or fabric. Bookbinders take this technique to a whole new level, though, producing incredible results. A marbling comb is one of the tools they can use to "rake" beautifully intricate patterns into the paints before transferring them onto paper.

The earliest form of marbling is believed to have started in Japan in the 12th century (although it's possible it began in China some two centuries before that). Legend has it that a chap called Jizemon Hiroba invented the process, which is known in Japanese as *suminagashi* (roughly translated as "floating ink"), after being inspired by the sight of the Kasuga Shrine – a Shinto shrine established in 768 CE in Nara, in south central Honshu.

Marbling is thought to have first been associated with paper crafting in the 15th century in Central Asia, but marbling combs were probably first used there at the end of the 16th century. Meanwhile, marbling spread to Europe, and it is thought that the close relationship between marbling and bookbinding (marbled paper being used for the endpapers of books) was established in France in the 17th century.

Marbling combs can be very simple tools and are often people make their own using long, thin pieces of wood with brass pins inserted at various intervals depending on the type of pattern you want to create.

CHAPTER 9
CLEANING & CARING FOR TOOLS

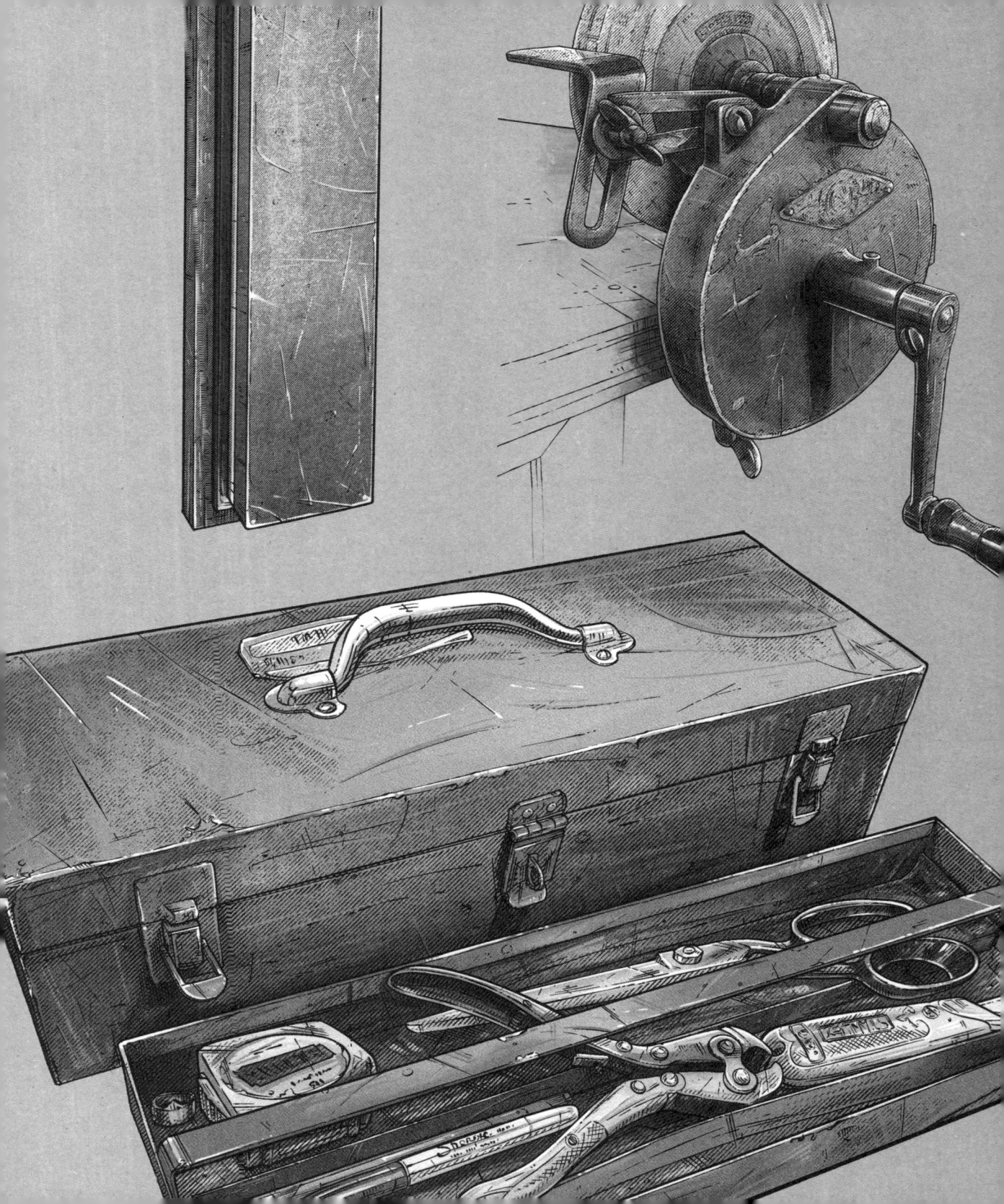

Tool No. 142

TOOLBOX

Before I went to college, I had a job as the Saturday boy at Karmann Konnection, a supplier of restoration parts and accessories for Volkswagen. I tried to learn as much as I could from John, the mechanic, who, every morning, would walk in with this red Snap-On toolbox.

My Snap-On toolbox is probably the only thing I've ever bought where I didn't try to haggle over the price.

He had other toolboxes, but this would be the first one out of the van in the morning before he made a coffee. It was crammed full and super heavy, and I just loved it and loved him. We became really good friends, and we still are now. He helped me with the Porsche episode on my YouTube channel, and now the tables have turned – he's working for me and he still walks into the workshop with that same toolbox.

Years and years ago, I was hunting through an auto jumble and stumbled across the same Snap-On toolbox. It even had the same sticker on the top. I couldn't believe it! Usually on the rare occasions you see one, it's in a terrible shape with no handle and massively bashed about. But this one was perfect – well-worn but lovingly looked after. It's probably the only thing I've ever bought where I didn't try to haggle over the price. I wanted it so much I think I would have paid whatever the bloke asked. You just can't buy them anymore, and I take it everywhere with me.

We did a reveal of a J40 pedal car at Goodwood the other week on *The Repair Shop* and I brought my toolbox along, because you never know when you might need it if a wheel comes loose or something. It reminds me of my childhood and John, and that's what set me on this direction doing what I do.

It wasn't my first toolbox; that was one of those ones with six folding compartments inside. It was grey and rubbish – I'd fill it with all sorts and it would be a jumble in there. With my red Snap-On toolbox, though, I find myself being careful about what I put in it. I don't want to fill it with rubbish. Soon as you have something like that, you just take care of it and put your best set of pliers in there. When you have special things to you, you want to look after them more. This toolbox is a special thing for me.

Tool No. 143

ENGINEER'S CABINET

Engineer's cabinets look like miniature chest of drawers with a handy shelf that pulls out underneath. I've got a collection of these and they're really special to me.

But my engineer's cabinets are not just for standing there looking pretty – there's a handle on top of them for a good reason! I use them all, carry them about with me and I know exactly what's in each one. Often they contain specialist tools that I've used on jobs for *The Repair Shop*, things that I'm working on back at the workshop, or commercial jobs I've had in the past. I've got one engineer's cabinet specifically for pigments and one with brushes for signwriting, for example.

Before the standard metal folding-out toolbox (everyone knows the one) was created, this would have been what a lot of craftworkers and apprentices used to carry around their tools. You might even find one in your grandad's shed somewhere. A lot of them feature the owner's initials either on the top of the box or on the lower drawer.

Some of the best British ones I've seen were made by Moore & Wright, a measuring tool company founded in Sheffield in 1906 by Frank Moore, a young engineer. The company was well-known for the quality of its components and attention to detail, and it obviously did a decent job of making engineer's cabinets, too.

Over in the US, H. Gerstner & Sons also made some truly beautiful engineer's cabinets. Harry Gerstner spent a year designing and making his first one in the evenings after he came back from his job working as a pattern maker. His mates and colleagues loved it, and soon he was being asked to make more. Eventually, he used a $100 bonus to start his own company, which he did in 1906. A bumper year for engineer's cabinets, clearly!

My engineer's cabinets are not just for standing there looking pretty – I use them all, carry them about with me and I know exactly what's in each one.

Bench grinder

Bench grinders are used mainly to sharpen blunt
tools. Electric grinders are available, but there's a
real satisfaction in using a hand tool for this job.

Tool No. 144

BENCH GRINDER

The bench grinder I often use is a classic hand-operated grinder that clamps to a workbench or table. It's the first thing that anyone visiting me in my little barn on the set of *The Repair Shop* turns, because it faces you as you come in and the handle is just so inviting. There is something magical about a hand-cranked grinder. It's something where the design has changed very little over centuries, so you feel that connection to the past when you're using one.

Bench grinders have a simple mechanism. The handle turns gears that rotate an abrasive wheel, usually made of sandstone, a hard sedimentary rock containing quartz or feldspar. I usually hold the tool in my left hand and crank the handle with my right. The wheel keeps going for a while after you've turned it a good few times, which always feels like you're freewheeling a bike.

I use a classic old bench grinder for sharpening the tungsten electrodes on my welder, but it's also great for sharpening tools such as chisels, drill bits and reshaping screwdrivers. I've got a Snap-On electric version if I need it, but I can't make too much noise working in my little barn at *The Repair Shop* when the cameras are rolling just outside, so I use hand tools whenever I can.

Using a tool like this all day would have been hard physical work, and it also needs a lot of concentration. That's why the phrase "nose to the grindstone" is thought to have come about. Tool sharpeners would have sat hunched over grinding stones, watching very closely in case the steel overheats. So they would have literally had their noses to the grindstone.

The phrase "nose to the grindstone" is thought to have come about because tool sharpeners would have sat hunched over grinding stones.

Tool No. 145

SHARPENING STONES

Also known as whetstones, sharpening stones have abrasive surfaces that are used to sharpen edge tools. They range widely in size, shape and the material they're made from, which can be natural or manufactured. Manufactured whetstones often have two faces with either fine, medium or coarse grit. You need to use a cutting fluid with some types of whetstone (which is why they are known as oilstones or waterstones) to help sharpen the surface and carry away the metal shavings.

One of the most famous naturally-occurring whetstones is coticule, a creamy-grey coloured sedimentary rock found in the Belgian Ardennes (a large area of forests and rolling hills, most of which is situated in southern Belgium). Coticule stones have been sought after since Roman times because of their impressive ability to sharpen blades.

But it wasn't until 1865 that someone made any real attempt to start a business selling them, and that was down to an enterprising lady called Denise Burton-Walrant. At first, she rented agricultural land and hired labourers to collect the stones, but when her son took over the business in 1901, they ramped up production, opened quarries and installed electrically-driven machines in a dedicated facility. The business grew and grew in the early 20th century, until the 1950s, when artificial whetstones became increasingly popular and the company couldn't compete. Production eventually stopped in 1982 after the owner died.

But that's not the end of the story, because Maurice Celis, a Belgian mining engineer, spotted an opportunity – he founded the company Ardennes-Coticule in 1998 and started quarrying the stones again. He also realized that the blue layer of rock found next to the coticule made for

another excellent whetstone. So he began to produce Belgian Blue stones as well, and went on a very successful campaign of going to professional fairs and talking to potential customers about the quality of the stones he produced. And so word began to spread.

Whetstones are also still quarried in Arkansas, USA, thanks to the beds of novaculite in the Ouachita Mountains (deposits have also been found in Japan, Syria and the Middle East). Novaculite is a dense, whitish-grey or black fine-grained sedimentary rock that has been used since prehistory to make arrowheads and blades before its sharpening qualities were discovered. The word "novaculite" comes from the Latin word *novacula*, which means "sharp knife" or "razor", and the stones were being used as hones for men's razors in Arkansas before the commercial value of them was exploited.

Japanese sharpening stones are traditionally submerged in water before you use them so the pores fill with water and create a smooth surface. Originally, naturally occurring whetstones were sourced in the area surrounding the city of Kyoto. Consisting of sedimentary rock containing very fine silicate grains and clay, the stones are softer than other sharpening stones, so they are not as aggressive on blades but they wear quickly so need to be looked after. Nowadays, Japanese waterstones are usually synthetic, but there are still some producers of impressive natural stone versions, although they aren't cheap.

INDEX

ACKNOWLEDGEMENTS

Making a book like this is a real team effort, and throughout the process I've been lucky enough to get support from some incredibly knowledgeable and talented people. I want to thank all of you who've helped me with my research and taken time to speak about some of the lovely old tools in my collection.

.

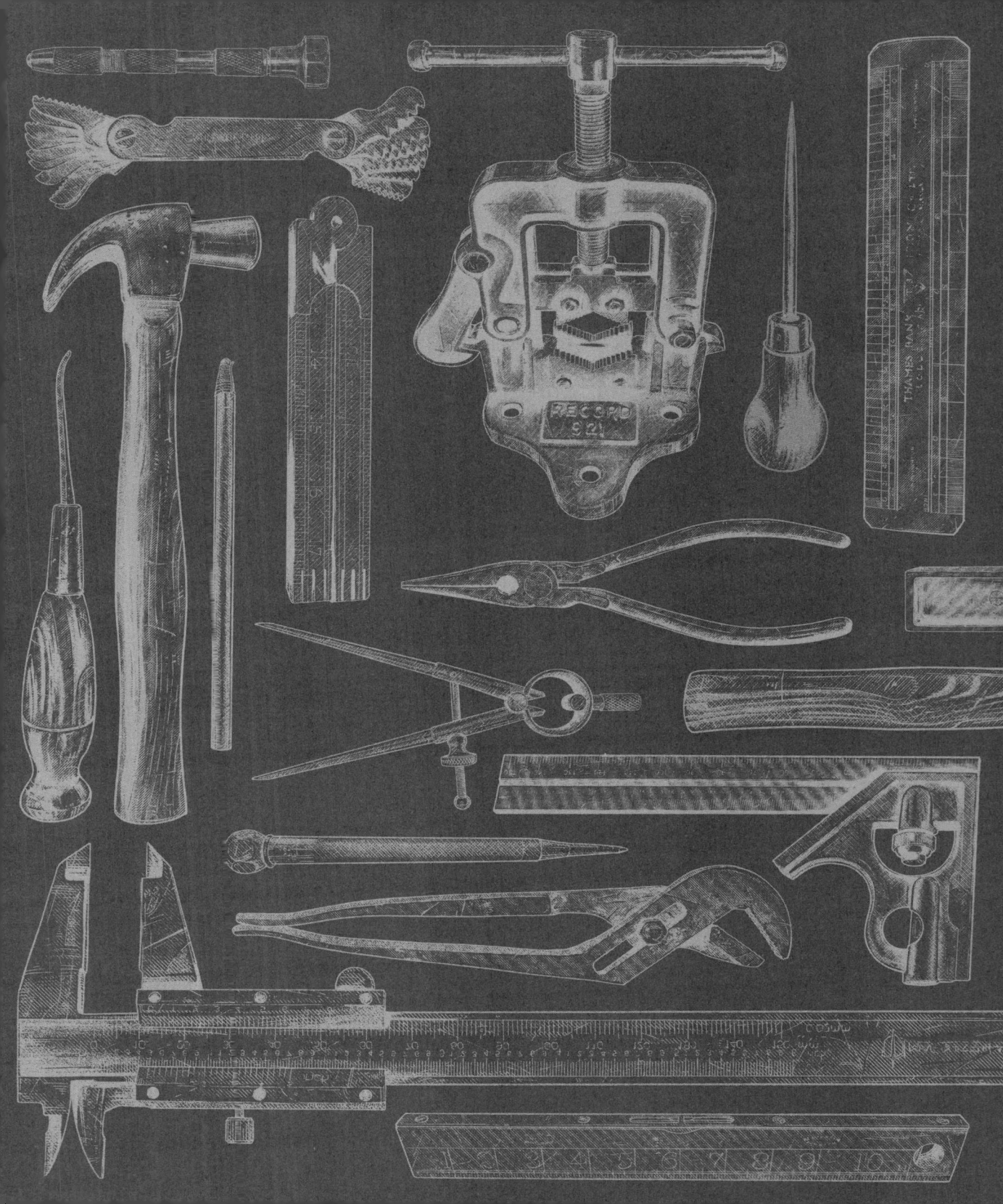